Plant Diseases

PLANT DISEASES

GOPAL SWARUP
D.R. DASGUPTA
P.K. KOSHY

ANMOL PUBLICATIONS PVT LTD
New Delhi-110 002

ANMOL PUBLICATIONS PVT. LTD.
Regd. Office: 4360/4, Ansari Road, Daryaganj,
New Delhi-110 002 (India)
Ph.: 23278000, 23261597
Branch Office: No. 1015, Ist Main Road, BSK IIIrd Stage
IIIrd Phase, IIIrd Block,
Bangalore-560 085 (India)
Tel.: 080-41723429
Visit us at: www.anmolpublications.com

Plant Diseases

Reprint, 2009

ISBN 81-7041-209-9

PRINTED IN INDIA

Printed at Mehra Offset Press, Delhi.

Preface

Research contribution on plant parasitic nematodes has increased tremendously during the last four decades or so and a vast volume of literature has been added. Till 1960, the major contributions were from the leaders in the area, U.S.A., U.K. and The Netherlands. However, after this period, nematological investigations were initiated on organised basis in Asian and African countries as well with post-graduate teaching programmes also. This is reflected in the number of nematological societies that have been formed, the latest being the Afro-Asian Society of Nematologists. Also, a number of scientific journals devoted exclusively to nematology have come into being catering for the research outlets. The first book on plant diseases caused by plant parasitic nematodes was published in 1951 by Christie, one of the pioneering scientists of the time. Eventhough, a large volume of literature has been added since then, knowledge about many nematodes is very limited. In a number of cases, informations are through trials on pathogenicity rather than any field disease problem. An attempt has been made, therefore, to bring together as much information as possible including effects on the plants in pathogenicity trials. Whenever considered feasible, some details on morphology and biology have been given for a better understanding of the nematode and the diseases caused by them. Whenever it was felt that informations with regard to disease aspects were rather limited, the nematode group has been treated as a whole rather than dealing with individual species. We hope that the book would be of use to the nematologists as well as the plant protection scientists.

We gratefully acknowledge the help rendered by Drs. Nandini Gokte, A.K. Ganguly, K.S. Varaprasad, V.K. Sosamma, Mr. Thakur Das and Mr. Gurbux Singh in the preparation of the manuscript.

We are also grateful to Dr. N.G.M. Hague, The University of California, Riverside, U.S.A. and Dr. M.S. Jairajpuri for permission to reproduce some photographs.

Gopal Swarup
D.R. Dasgupta
P.K. Koshy

Contents

1

Introduction

Nematology is a recognised discipline now in almost all the countries of the world. In its infancy or in some instances even now, it has not generated any significant interest amongst zoologists, in spite of the fact that more than 10,000 species have been described—no doubt only a fraction of the actually existing ones. At the same time, botanists as well as plant pathologists have kept away from this area because of the animal nature of the nematodes. However, with more and more informations accumulating, regarding their role in plant diseases, either singly or in complex associations with other soil micro-organisms, like fungi and bacteria, as also the role of some of the nematode species in vectoring soil-borne viruses, the science of nematology, to be more precise, agro-nematology, has stimulated interests of the agricultural scientists. The unique nature of the plant parasitic nematodes, their ubiquitousness and persistence as soil-borne plant pathogens, the unobtrusiveness of their attack, the tremendous losses they have caused to some of the economically important agricultural crops, have further established this group of pests on par with other allied disciplines.

In spite of the tremendous importance of the science of agro-nematology, several countries of the world still lack in the knowledge of various types of nematodes present in soil, their role and behaviour, frequency, distribution, etc. The extensive as well as intensive works reported from U.S.A., England and The Netherlands, the leaders in this branch of Science, have geared the thinking of agricultural scientists to this field in several countries of the world, particularly the developing countries. With the vast increase in exchange and distribution of plant material, during modern times, it has brought forth a necessity of technical knowhow of looking for plant nematodes so that the noxious types are not introduced in countries where the particular types are not recorded to exist. The importance of nematodes has been beautifully expressed by Cobb, father of modern nematology, as '...if all the matter in universe except the nematodes were swept away, our world would still be dimly recognisable.. we would find its mountains, hills, valleys, rivers, lakes, and oceans represented by a film of nematodes'[1].

Plant nematology forms a link between zoology and botany as its study involves not only the morphology and taxonomy of the animal but also their biology, host-parasite relations, control, etc. To have proficiency in the subject, therefore, one has to have knowledge not only of plant anatomy, plant pathology, physiology but also of zoology and parasitology. The plant diseases caused by nematodes, symptoms, identification of the causal nematodes, etc., therefore, form an important aspect in the overall understanding of the subject.

Plant diseases have two important aspects, economic and ecologic. One or both of these aspects may cause a tremendous effect on the development of the discipline. Identification of the presence of the golden nematode of potato in Long Islands (U.S.A.), bulb and stem nematodes in Europe, 'molya' disease of wheat and barley (cereal root eelworm) in India are some of the outstanding examples where nematodes have caused economic as well as ecological impact on the scientists as well as the administrators. Several disease problems still remain unrecognised or have not been able to attract attention because of the limited expertise available or uneconomic crop plants attacked by them. The subterranean habit of most nematodes, their microscopic size, lack of any clear cut above ground symptoms, the slow debilitating type of damage, rather than a killing effect on the host, want of any diagnostic service to provide advice on the basis of soil and plant tests, etc., are some of the important reasons which have precluded any sustained attentions from the farmers and administrators. Furthermore, the plants are rarely subject to association with only one pathogenic organisms. There is a growing amount of evidence that nematodes, in quite a few cases, particularly in the case of soil-borne diseases, may act as initiators, thus facilitating parasitism by weak pathogens. Indications to their role as aggravators, resistance breakers and deterrents of plant diseases are now well recognised as well as understood. In these complex associations, therefore, the disease syndrome is influenced by two or more concurrent pathogenic organisms. Such associations may be more common in nature than is being realised at the present juncture. The total disease picture, therefore, necessitates a thorough understanding of the influencing factors of the various organisms in the symptom expression of the diseases.

A very commendable effort was made for the first time by Christie[2] when he compiled all the available informations in his book 'Plant Nematodes, their Bionomics and Control'. More than thirty years have elapsed since its publication. A large volume of literature has been since added and several new disease problems have come to the forefront. However, in spite of intensified researches, the literature is mainly confined to well known nematode problems only. Informations on ectoparasites or nematodes like the lance or spiral nematodes, which are the most common as well as often preponderant populations in agricultural soils, are still meagre. In the present compilation, an attempt has been made to bring together as much uptodate information as possible for the students of nematology.

REFERENCES

1. Thorne, G., Principles of Nematology, McGraw-Hill Book Co., New York, 553 pp., 1961.

2. Christie, J.R., Plant nematodes, their bionomics and control, Florida Agricultural Experiment Station, Gainesville, 256 pp., 1969.

2

Historical Aspects

The knowledge of namatodes can be traced to pre-Christ era. Mention of certain epithets in Indian scriptures (4000-6000 B.C.) and Egyptian records (1553 B.C.) indicate a possible awareness of atleast animal nematodes. Lacking the microscopes in the early periods, larger animals, visible to the naked eye, were naturally first recorded and investigated. These large forms were parasites of animals and human beings. The Guinea worm, *Dracunculus medinensis*, causing severe pain and swelling in arms and legs of humans, and the round worm, *Ascaris lumbricoides*, appear in the early Egyptian writings. Aristotle (384-322 B.C.) in his articles has also made frequent references to animal p rasitic nematodes. Comparatively, informations on plant parasitic nematodes are of recent origin.

It is not unlikely that the free living nematodes developed and increased with the first tilling of the land by man. It has been speculated that the 'soil exhaustion' theory, debated in Athens about 2,500 years ago and which might have been one of the factors in disappearance of certain civilizations, could well be as a result of 'increase of soil borne pathogens of all kinds' and due to climatic changes or lack of fertilizers, since the Romans are considered to have had high knowledge of agricultural technology. The current knowledge about the sugarbeet sickness in Europe or the yellows disease of pepper in Indonesia or the slow decline of citrus in California (U.S.A.), to mention a few, long considered due to soil disorders or deficiency and now established as nematode problems, lend support to the suggestion that nematodes may have been one of the important contributors to the 'soil exhaustion' during the Roman period.

Shakespeare's mention of 'sowed cockle, reap'd no corn' in his drama Love's Labour's Lost in 1594 could be the first mention of a plant parasitic nematode. However, during the post-Christ era, the first published record is that of a free living nematode, *Turbatrix aceti* in 1656[1]. Almost 94 years later, Needham, a clergyman discovered the first plant parasitic nematode when in 1743[2], he recorded the worms present inside galls or 'cockles' of wheat. He described these worms in the following words: 'Upon opening the small black grains of smutty wheat, which they here distinguish from the blighted corn, the latter affording nothing but black dust, into which the whole substance of ear is converted, I perceived a soft fibrous substance, a small portion of which I placed on my object plate. It seemed to consist wholly of longitudinal fibres bundled together; and you will be surprised, perhaps that I should say, without any least sign of life or motion. I dropped a globule of water on it in order to try, if the parts when separated might be viewed more conveniently when to my great surprise these imaginary fibres as it were, separated from each other, took life, moved regularly not with a progressive but twisted motion and continued to do so for a space of 9 to 10 hours, when I threw them away. I am satisfied that they are the species of aquatic animals and may be denominated worms, eels, or serpents, which they very much resemble'.

Vibrio tritici was the name given for the ear-cockle nematode in 1799[3]. Later, it was reported from some grasses under the name of *V. agrostis*. Another species, *V. graminis*, was added to the literature when it was found to be present on *Fescue* grass[4]. Davaine[5] in 1857 gave a detailed description of its life cycle. The root-knot nematodes were the next plant parasitic nematodes discovered in 1855[6] on greenhouse cucumbers. The disease was described as producing galls on roots and containing eggs and 'vibrios'. This was closely followed by the record of *Anguillula dipsaci* (*Ditylenchus dipsaci*) infesting teasel, *Dipsacus fullonum*, in 1857[7]. Not long after these reports, the sugarbeet cyst nematode was discovered in 1859.[8] More than ten years elapsed before the nematode was studied and named as *Heterodera schachtii*[9]. The sugarbee industry was very important, at this time, and for this reason, the discovery of this nematode problem attracted lot of attention. It was found that this nematode was responsible for the prevailing 'beet sickness' problem. A number of scientists got interested in the problem and put forth a number of publications amongst which there was a classic contribution on the life cycle of the nematode[10]. The other important plant parasitic nematodes discovered during the 19th century were: *Heterodera goettingiana* on pea roots[11] ; *Ditylenchus radicicola on Poa annua* and other grasses ; and the foliar nematode, *Aphelenchoides fragariae*, on strawberry plants[12].

The 19th century was marked also with the first attempt at controlling the nematodes with the use of chemicals and one of the first was the use of carbon di sulfide against sugarbeet cyst nematode[13].

Besides the discovery of important plant parasitic nematodes in this century, the period was marked also with important taxonomical contributions, notably by Dujardin[14] in 1845 on *Dorylaimus* and *Rhabditis*; monographs by Bastian[15] in 1865 and Butschli[16] in 1873; the classical monograph by deMan[17] in 1884 entitled 'Die einheimschen, frei in der rainen Erde und im sussen-Wasser lebenden Nematoden der Niederlandischen Fauna' which, to this day, has stood the test of changing concepts and still ranks as one of the finest contributions to the science of nematology. Another important name needing mention is thal of Oerley[18] who in 1880 contributed a monograph on Anguillulidae, with illustrations and informations on 202 species belonging to 27 genera of free living and plant parasitic nematodes.

Besides *Heterodera*, *Ditylenchus* and *Aphelenchoides*, the other plant parasitic nematodes described during the 19th century were *Halenchus*, *Hoplolaimus*, *Longidorus*, *Paratylenchus*, *Radopholus*, *Rotylenchus*, *Trichodorus* and *Xiphinema*.

Nematology in the 20th Century. The progress of nematology in this period can be best described in two phases : (1) the pre-World War II period ; and (2) the post-World War II period. The first period is dominated by the personality of Nathan Augustus Cobb, who after a chequered career joined the United States Department of Agriculture in Washington and from the year 1915 till his death in 1932 devoted his full energies to the study of the science of nematology. The coining of the word 'nematology' is credited to him and his compilation 'Contributions to a Science of Nematology'[19] is a tribute to his knowledge and mastery of the subject. He was instrumental in evolving many of the techniques like wet screeing to separate nematodes from soil, methods of preserving and mounting nematodes, section cutting, etc., which are still in vogue today. He is considered as the Father of modern nematology and was the man behind the establishment of the Division of Nematology at the United States Department of Agriculture.

Almost simultaneously with the contributions of Cobb in the U.S.A., Micoletzky in Austria started compiling and bringing together published works on soil and fresh water nematodes which he published as 'Die freilebenden Erd-Nematoden' in 1922[20]. In this compilation, he presented keys for identification of the species. At the same time, Tom Goodey in England and I.N. Filipjev in Russia made tremendous impacts on the development of nematology in their respective countries. Tom Goodey, besides being the Head of the Department of Nematology at the Rothamsted Experiment Station, U.K., published two books : (1) Plant parasitic nematodes and the diseases they cause, 1933 ; and (2) Soil and fresh water nematodes, 1951, which proved invaluable aids in the study of the subject. The former book dealt with the disease aspect while the latter contribution was mainly of taxonomic nature. He published also a number of scientific papers based on his research work.

Filipjev was equally active in his country and besides contributing to the taxonomic concepts, wrote a book in Russian which was later translated into English by his colleague J.H. Schuurmans-Stekhoven under the title of 'A manual of Agricultural Helminthology' with the publication date as 1941.

Other contemporary workers of repute during this period were : B.G. Chitwood, A.L. Taylor, J.R. Christie, G. Thorne in U.S.A. ; M.T. Fraklin, B.G. Peters, D.W. Fenwick in U.K. and Marcinowski and Goffart in Europe.

Several new plant parasitic nematodes were discovered and added to the growing list of nematode species. Thorne, as Head of the Division of Nematology at the United States Department of Agriculture in U.S.A., initiated the concept of maintaining catalogued collections and specimens for ready use of all those interested in the science. By the time he retired in 1956, he had built up a valuable collection of more than 36,000 specimens.

Post-World War II Period: The period upto pre-World War II was marked with descriptions of new species, identification of new nematode problems and the nematode species associated with them, development of new techniques and taxonomic concepts and writing of books and monographs. No attempt could be made to achieve field control of nematode problems since there was no inexpensive chemical that could be used profitably in any control programme. In this respect, the post-World War II period is outstanding for the almost simultaneous discovery of dichloropropene+dichloropropane (D-D) mixture by Carter in 1943[21] and ethylene dibromide (EDB) by Christile in 1945[22]. These two soil fumigants were not only very effective against the nematodes but also were comparatively inexpensive to apply. These nematicides replaced all the earlier chemicals like chloropicrin and carbon di sulfide which were expensive as well as highly toxic and could be used profitably under greenhouse conditions only. Besides nematicidal in value, DD and EDB had low toxicity to fungi and bacteria and therefore, these proved to be highly useful in demonstrating the effects of nematodes on plants.

Despite the revolutionary era started by the discovery of these chemicals, proof of nematode pathogenicity and fulfilling all the Koch's postulates was not achieved so far. Even as early as 1914, Byars[23] had developed techniques to study life cycle of nematode (root-knot) under sterile tomato culture. However, this aspect was not studied seriously till 1955 when Mountain[24] used tissue culture technique to obtain sterile cultures of nematodes. He succeeded in culturing *Pratylenchus minyus* on excised maize roots, growing in sterile nutrient agar. The technique was useful in demonstrating development and production of root damage by *P. minyus* on seedlings growing under sterile conditions. At a later date (1959), Mountain and Patrick[25] were able to obtain sterile *P. penetrans* population by treating the nematodes with bactericide and fungicide solutions. These nematodes were then inoculated to sterile peach seedlings, growing in tubes, under sterile conditions. Production of lesions on the roots was recorded within 90 minutes of inoculations and isolations from these lesioned areas failed to show any bacterial or fungal infections and yielded only nematode populations. It was, therefore, demonstrated that *P. penetrans* could invade and kill peach root tissues in the absence of bacteria or fungi.

About this time, another landmark was set up by Christie and Perry[26] when in 1951, they demonstrated conclusively that *Trichodorus*, an ectoparasite, was a serious problem of crops in Florida, U.S.A. This opened up the path for detailed studies on other ectoparasitic nematodes which had been neglected hitherto for want of informations and techniques to study them.

These researches opened up great possibilities and avenues in the study of nematodes and a number of new findings and discoveries helped in the rapid strides in advancing our knowledge of the subject as well as in initiating worldwide interest and activities. The discovery of golden nematode infestation in potato in Nassau County, Long Islands, U.S.A. and the destruction of 20 million pepper trees in Indonesia, primarily by *Radopholus similis*, activated various agencies of the state and federal governments. The need for trained personnel stimulated organisation of workshops and training programmes resulting into much publicity about plant parasitic nematodes as important factors in limiting agricultural production. Thus, new job opportunities were provided to the agricultural scientists.

In the late fifties, an outstanding discovery was made about the role of nematodes in transmitting virus diseases. It was demonstrated that the fanleaf virus of grapevine is transmitted by the nematode, *Xiphinema index*[27]. This added new dimensions to the investigations on the nematode-micro-organism complexes. Several virologists, battling with the soil borne viruses, found the breakthrough in the problem with more and more evidences of nematodes acting as vectors for such viruses.

The history of scientific advancement in nematology would be incomplete without the mention of the name of Myuge who in 1956[28] demonstrated that there are more proteolyctic substances in the galled root-knot tissues than in the healthy tissues. He demonstrated also that alcohol extracts of the galled tissues can induce galling in healthy tissues. Earlier, Linford[29] in 1937 had shown that the root-knot larvae inject salivary juices in the host tissues while feeding but this was not seriously taken up till Miyuge's work provided the required stimulation to the nematologists.

Historically, starting of formal lectures and practical classes at the University of California, U.S.A. in 1948 assured the future of nematology in relation to other allied disciplines. Subsequently, several other universities in the U.S.A. included study of nematology in their curriculum.

By 1955, nematologists or workers interested in the subject had grown in numbers, large enough for an organisation of a scientific society in Europe. It was first named as the Society of European Nematologists but with more and more people, from other parts of the world exhibiting an interest in joining the Society, the name was changed to European Society of Nematologists. The activities of the Society started with the issue of newsletters and subsequently a full fledged scientific journal under the name of Nematologica was started in 1956. It is published quarterly from Leid n, The Netherlands. The Society of Nematologists (SON) in the U.S.A. was the next one to be organised in 1961. It also started with a newsletter and later a journal under the name of Journal of Nematology was put forward by the Society in 1969. The Nematological Society of India was founded in 1969 followed by publication of its journal from 1971 under the name of Indian Journal of Nematology. The organisation of Tropical American Nematologists (OTAN) came into being in 1971 and a society was also formed in Japan in 1972. The latest in the series is the formation of Afro-Asian Society of Nematologists (AASN) in 1984. These developments indicate the rapid strides the science has made within the last 25 years or so.

Amongst the Middle East, African and Asian countries, India has emerged as one of the most important research centre for the nematological investigations. Therefore, a brief mention is being made here about the developments of nematological research in the country.

Nematology in India. Frequent references to human parasitic nematodes can be found in the ancient Indian scriptures (Vedas, 6000 4000 B C.) under the Sanskrit name 'Krimin'. According to these books, 'Krimis' (vernacular name for nematodes) were snake-like worms infesting human intestines and being responsible for such complications as diarrhoea, feverishness, itching, etc. In these writings, the 'Krimis' have been divided into two broad groups : (1) visible (the Dristam) ; and (2) the invisible (Adrishtam). These organisms have been recorded to be prevailing on hills, jungles, in medicinal herbs, animals, drinking water, etc.

The first authentic report, however, of the presence of any plant parasitic nematodes in the country was made by Barber[30] in 1901 when he described root-knot nematodes on tea (*Camelia sinensis*) under the then prevailing name *Heterodera radicicola.* Between 1901 and 1945, occasional reports appeared recording occurrences of root-knot as well as other parasitic forms. The main contributors during this period were Ayyar, Baylis and Mayne and Subramaniam. A sustained research tempo was built up only in the late fifties with the scientists at the Indian Agricultural Research Institute, New Delhi (IARI) ; Aligarh Muslim University, Aligarh and Osmania University, Hyderabad, taking lead in conducting surveys which yielded valuable informations about the type of nematodes and their distribution in Indian soils. These activities led to the formation of organised laboratories at various other agricultural universities for whole time research in nematology.

One of the most important milestones in the development of nematology in India was identification of association of cereal root eelworm, *Heterodera avenae*, with a wheat disease, locally known as *Molya* in the north-western part of the country. It was soon followed by the discovery of golden nematode of potato in the southern part of the country[31]. These two nematode problems posed a serious threat to the cultivation of the respective crops, resulting into an immediate interest and awareness amongst the administrators for the need to organise nematological research throughout the country. Sections of nematology were added in the departments of plant pathology and entomology of the Indian Agricultural Research Institute, New Delhi (IARI) and also at the Agricultural Research Institute, Coimbatore. At this stage, the Rockefellor Foundation in India stepped in and arranged consultancy services of Professor, D.J. Raski of the University of California, U.S.A. for IARI, New Delhi. This helped in getting equipments and training of the scientists. In 1964, the first training course in Nematology was organised under the expert supervisions of three eminent scientists, Late Dr. J.B. Goodey and Dr. F.G.W. Jones from U.K. and Dr. D.J. Raski from U.S.A. Soon after this, a series of seven training courses were organised, with the collaboration of the Netherlands government, at IARI, New Delhi and Aligarh Muslim University, under the name of South-East Asia Nematology Training Course.

The importance of nematodes to agriculture was fully appreciated now and full department status was granted to nematology at the IARI, New Delhi. India took the lead in world by starting the first post-graduate courses in the discipline, leading to the award of Ph.D. and M.Sc. degrees. The other agricultural universities of the country followed this lead.

The need for evaluation of chemical control, under Indian conditions, was felt and accordingly a FAO sponsored project was started under the expert supervision of Dr. D.J. Raski of U.S.A. It was possible to demonstrate a 4-fold increase in yield of wheat and a 16-fold increase in barley yield in fields infested with 'molya' disease with the use of DD. Similar trials against other nematode problems and crops proved very demonstrative, particularly against sugarcane.

A similar Indo-German collaborative programme was developed for the golden nematode of potato. These trials were very helpful in bringing about an awareness amongst the cultivators about the nematode problems.

The future of the nematological research and nematologists was assured with formation of the Nematological Society of India in 1969 and the publication of the Indian Journal of Nematology from 1971. The IARI, New Delhi has set up also a National Nematode Collection which has become one of the important centres for deposition of tppe slides.

REFERENCES

1. Borellus, P. Historiaruum, et observationum medicophysicarum, Centuriae quattuor, 384 pp., 1656.
2. Needham, T., A letter concerning certain chalky tubulous concretions called malm ; with some microscopical observations on the farina of the red lily, and of worms discovered in smutty corn, *Philos. Trans. Roy. Soc. 42*, 634-641, 1743.
3. Steinbuch, J.G. Das Grasalchen, *Vibrio agrostis*, *Naturforscher*, *28*, 223-259, 1799.
4. Hardy, J., On the effects produced by some insects, etc., on plants, *Ann. Mag. Nat. Hist.*, 6 (2nd ser.), 182-183, 1850.
5. Davaine, C. Rechereches sur 1' anguillule du ble nielle consideree au point de vue de 1' histoire naturelle et de 1 'agriculture, Paris, 88 pp., 1857.
6. Berkeley, M.J. Vibrio forming excrescences on the roots of cucumber plants, *Gard. Chron. 14*, 220, 1855.
7. Kuhn, J. Uber das Vorkommen von *Anguillula* an erkrankten Bluhten-koplen von *Dipsacus fullonum*, *Zeitschr. wiss. Zool. 9* 129-137, 1857.
8. Schacht, H. Ueber einige Feinde und Krankheiten der Zuckerrube. *Zeitschr. Ver. Rubenzucker'Ind. Zoolver. 9*, 390, 1859.

9. Schmidt, A., Ueber den Ruben-Nematoden (*Heterodera schachtii*). *Zeitschr. Ver. Rubenzucker-Ind. Zoolver 22*, 67-75, 1871.

10. Strubell, A., Untersunchungen uber den Bau and die Entwicklung des Ruben-Nematoden, *Heterodera schachtii. Bibliothec Zool. 2*, 1-52, 1888.

11. Liebscher, G., Beobachtungen uber des Auftreten eines Nematoden an Erbsen, *J. Landwirts, 40*, 357-368, 1892.

12. Ritzema Bos, J., Zwei neue Nematodenkrankheiten der Erdbeer pflanze, *Zeitschr, Pflanzenkrank*, 1, 1-16, 1891.

13. Kuhn, J., Die Ergebnisse der Versuche zur Ermittelung der Ursache der Rubenmudigkeit und zur Erforschung der Natur der Nematoden. *Ber. Physiol. Lab. Univ. Halle. 3*, 1-153, 1881.

14. Dujardin, F., Histoire naturelle des helminthes, Paris, 1845.

15. Bastian, H.C., Monograph on the Anguillulidae, *Linn. Soc. Lond. Trans. 25*, 73-184, 1865.

16. Butschli, O. Beitrage zur kenntnis der freilebenden Nematoden, *Nova Acta Acad. Leop. Carol. 36*, 1-124, 1873.

17. Man, J.G. de-Die frei in der reinen Erde und im sussen-Wasser lebenden Nematoden der Niederlandischen Fauna, 206 pp., 1884.

18. Orley, L., Az anguillulidak maganrajza, Monographie der Anguilluliden. Budapest, 165 pp., 1880.

19. Cobb, N.A., Contributions to a science of nematology, Pt. I, 1-33, 1914.

20. Micoletzky, H., Die freilebenden Erd-Nematoden, *Arch. Naturgesch*, Berlin, 87, 1-650, 1922.

21. Carter, W., A promising new soil amendment and disinfectant, *Science*, 97, 383-384, 1943.

22. Christie, J.R., Some preliminary tests to determine the efficacy of certain substances when used as soil fumigants to control the

root-knot, nematodes. *Proc. Helminth. Soc. Wash.*, 12, 14-19, 1945.

23. Byars, L.P., Preliminary notes on the cultivation of the plant parasitic nematode, *Heterodera rodicicola*, *Phytopathology*, 4, 323-326, 1914.
24. Mountain, W.S., A method of culturing plant parasitic nematodes under sterile conditions, *Proc. Helminth. Soc. Wash.*, 22, 49-52, 1955.
25. Mountain, W.S. and Patrick, Z.A., The peach replant problem in Ontario, VII, The pathogenicity of *Pratylenchus penetrans*, *Can. J. Bot.*, 37, 459-470, 1959.
26. Christie, J.R. and Perry, V.G., A root disease of plants caused by a nematode of the genus *Trichodorus*, *Science*, 113, 401-403, 1951.
27. Hewett, W.S., Raski, D.J. and Goheen, A.C., Nematode vector of soil-borne fanleaf virus of grapevines. *Phytopathology*, 48, 586-595, 1958.
28. Myuge, S.G., The tropic characteristic of the root-knot nematode, *J. Gen. Biol.* (USSR), 17, 1-21, 1956.
29. Linford, M.B., The feeding of the root-knot nematode in root tissue and nutrient solution, *Phytopathology, 27*, 824-835, 1937.
30. Barber, C.A., A tea eelworm disease in south India, *Dept. Land Records and Agric.* Bull. 45, 1901.
31. Jones, F.G.W., The Potato root eelworm, *Heterodera rostochiensis* in India, *Curr. Sci. 30*, 187, 1961.

3

Importance of Plant Parasitic Nematodes in Agriculture

Plant parasitic nematodes are unique in their ubiquitousness and persistence as soil borne pathogens. The unobtrusiveness of their attack allows their presence to often pass unnoticed while crops slowly decline in vigour and yield. Rarely is any crop free from attack of these tiny and microscopic pathogens. They are present every where,. in fields, glasshouses. orchards, home gardens, and yet their presence is generally not felt till the concerned individual is baffled by the continuous decline of the crop in spite of best agronomic practices. The losses, as a result of nematode attack do not consist of yield reductions only but other aspects like lesser ability of infected roots to utilize fully the available nutrients from soil or the necessity to grow uneconomical rotational crops, in an effort to control the noxious nematodes, are some of the effects which follow as a result of nematode infestation. Considerable information have been generated on the losses caused by the nematodes to staple and other crops, particularly in U.S.A., U.K. and the Netherlands.

Amongst the soil inhabitants, nematodes constitute one of the most important groups of organisms which quite often play a vital role in the deterioration of the health of the standing crops. With an intensification in agriculture and also with continued and uninterrupted new introductions in countries, it is obvious that there is every likelihood of hitherto unknown problems which may become serious in the near future. The golden nematode problem in Long Islands, U.S.A. and in the southern part of India is one of the most outstanding examples of such type.

There are still no reliable information on the exact losses caused by the nematodes to various crops. Most of the reports are approximations based on indirect evidences, interpreted on the basis of pot trials and chemical trials under field conditions. The validity of the results on yield increases, as a result of chemical applications, is open to question since the total effect on biological flora of the soil is not fully understood and therefore, it might not be justifiable to interpret results of field trials with nematode damage alone. The quantitative aspect of the nematode damage to the crops is not only affected by the size of the nematode population but also on other variables such as the species of the host plant, temperature and other environmental conditions, the biological flora and fauna of the soil, which singly or jointly may contribute in the overall expression of disease syndrome. A range from death of seedlings/plants, in some cases, to an absence of recognisable damage may not be uncommon in nematode infestations. One of the most important factors influencing yield of crops is the size of the nematode population in soil and also availability of a susceptible host. Measureable damage occurs only when the population density exceeds a certain limit. Experiments have demonstrated that about 100 nematodes (*Tylenchorhynchus claytoni*) can reduce the top growth of inoculated tobacco plants[1]. On the other hand, another species of the same genus (*T. martini*) causes deterioration of sugarcane roots at a population between 600 to 6400[2]. In onions, only 5 individuals of *Pratylenchus penetrans* per seedling are reported to cause serious damage.

Amongst the best known plant parasitic nematodes are the root-knot nematodes, beloging to the genus *Meloidogyne*, which have a wide host range and are widely prevalent all over the world. These produce conspicuous galls on the roots and the infestation can be recognised easily in fields. The wide distribution of these nematodes through tropics, sub-tropics and temperate regions and their occurrences in cultivated fields and greenhouses as well, make them as one of the most common agricultural pests. It is specially a serious problem with vegetable growers as also to cotton and tobacco growers. In majority of the cases, soil micro-organisms like fungi and bacteria are also associated with root-knot nematode infestations. Plants resistant against fungal and bacterial pathogens may become susceptible in the presence of nematode infestation. In some instances, root injuries caused by nematode larval penetration or feeding may provide an easy entry avenue for fungal and bacterial colonisation, which otherwise they might not have been able to do so independently. In such cases the disease expression may be quite severe.

Involvement of various ecto-parasitic nematodes, that feed ectoparasitically on plant roots, besides the other endo or semi-endoparasites, notably *Patylenchus* species (lesion nematodes), *Tylenchulus semipenetrans* (citrus nematode) and *Rotylenchulus reniformis* (reniform nematode) in the plant disease, adds to the dimensions of losses caused by these noxious nematodes to economically important crop plants. As stated earlier, many of these may not cause conspicuous damage immediately but with continued root feeding, the root growth may stop and the affected plants may express unthriftiness and stunted growth. However, not all the soil nematodes are harmful to the plants. Some of dorylaims and mononchs prey on nematodes, rotifers and soil micro-organisms. These and some predacious fungi, which attack nematodes, may well be maintaining a natural balance in the soil fauna in favour of the plant growth.

Even though a very large number of plant parasitic species have been recorded and documented, only 15 have been identified, so far, as world problems (Table 1). Besides these major nematode problems, there are several other important nematode diseases which cause heavy losses in certain parts of the world and under specific conditions. For example, the ear-cockle disease of wheat (*Anguina tritici*) could lead to as high as 80 per cent or more crop loss in certain pockets of India, if contaminated seed is regularly used for sowing. The maize cyst nematode (*Heterodera zeae*) is another nematode problem which has gained an economic aspect recently in India as well as U.S.A. The 'ufra' and white tip diseases of rice, respectively caused by *Ditylenchns angustus* and *Aphelenchoides besseyi*, often are responsible for serious crop losses in south-east Asian countries. Similarly, the soybean cyst nematode, *Heterodera glycines*, is of great economic importance in U.S.A. and has necessitated development of intensive nematode management programmes. The lesion nematodes, in general, and particularly *Pratylenchus coffeee* on coffee in southern parts of India and *P. loosi* on tea in Ceylon (Sri Lanka) have attracted great deal of attention from the growers as well as the scientists. The red ring disease of coconut caused by *Rhadinaphelenchus cocophilus* in Surinam and British Guiana and on oil palm in Venezuela is rated as one of the most devastating disease throughout the world.

There are several other known examples where nematodes have been established definitely as the causal organisms of the well known maladies. However, many are still unexplored but at the same time, the importance of nematodes, as one of the important limiting factors in crop production is now well established. Nematode control could play a primary role in reducing the existing gap between food and population.

TABLE 1

World Nematode Problems of High Economic Importance

Nematode	Year of First Discovery	Distribution	Crops Attacked	Economic Damage (%)
(1)	(2)	(3)	(4)	(5)
Root-knot nematodes (*Meloidogyne species*)	1855	Throughout World (England*)	Vegetables, Pulses, Oil-seeds Fruits, Fibre, Plantation and Few Cereal Crops	4 on[4] tomato[3]
1. *M. incognita*) 2. *M. javanica*) . *M. hapla*) 4. *M. arenaria*)				

(Contd.)

(Contd.)

(1)	(2)	(3)	(4)	(5)
5. Reniform nematode (*Rotylenchulus reniformis*)	1940	Tropical and semi-tropical areas of world (USA*)	Vegetables, Pulses, Oilseeds, Some Fruits and Fibre Crops	Cotton and Soybean (USA), Vegetables (Puerto Rico), Pulses (India)
Potato cyst nematodes (*Globodera spp.*)	1881	48 Potato growing	Potato, Tomato, Brinjal	10 (USA)[5] 9 (UK)[5]
6. *G. Rostochiensis*)		countries		
7. *G. Pallida*)		(Europe, America, Asia) (Germany*)		
8. Stem and bulb nematode (*Ditylenchus Dipsaci*)	1857	Germany*, USA, UK, France, USSR, Netherlands	300 plants including Vegelables, Cereals and Ornamentals	60 on Onion and Garlic[6] 18 on Corn[7]
9. Cereal root eelworm (*Heterodera avenae*)	1874	31 Wheat growing countries (Germany*)	Wheat, Barley, Oats, Rye, Maize	73-89[8] 4 Million Dollars-India
10. Sugarbeet cyst nematode (*Hetrodera schachtii*	1859	40 Countries including European	218 species belonging to 23 plant families	10—Sugarbeet[9], 25—Central

(Contd.)

(*Contd.*)

(1)	(2)	(3)	(4)	(5)
		(Germany*), American, African and Middle East	Including Sugar-beet	Europe[10]
11. Burrowing nematodes *Radophclus*) *spp.*) *R. similis*	1891	Tropical and sub-tropical parts of world (Fiji*)	250 plants including Banana, Pepper, Coconut	—
12. *R. citrophilus*	1928	USA*	1275 species and Relatives of Citrus	40-80 on Grapefruit and Oranges[11]
13. Citrus nematode (*Tylenchulus semipenetrans*)	1889	Throughout world (USA*)	75 Rutaceous plants including citrus and 9 non-Rutaceous plants	8.7-12.2[12]
14. Rice root nematodes *Hirschmanniella species*)	1902	46 Rice growing countries (Indonesia*)	Rice	25[13]
15. Pine wood nematode (*Bursaphelenchus xylophilus*)	1971	Japan*, USA, France	Pine	2,400,000m[3] of timber annually[14]

*Country of first record.

REFERENCES

1. Graham, T.W., The tobacco stunt nematode in North Carolina, *Phytopathology*, 44, 12-13, 1954.

2. Birchfield, W. and Martin, W.J., Pathogenicity on sugarcane and host plant studies of a species of *Tylenchorhynchus*, *Phytopathology*, 46, 277-280, 1956.

3. Valdez, R,B., Nematodes attacking tomato and their control, *First Int. Symp. on Tropical Tomato*. Asian Vegetable Research Centre, Taiwan, Ed. R. Cowell, 290 pp., 1978.

4. Brodie, B.B., Nematode parasites of potato, In Plant and Insect Nematodes (Ed. W.R. Nickle), Marcel Dekker Inc., New York, 167-212, 1984.

5. Evans, K. and Stone, A.R., A review of the distribution and the biology of the potato cyst nematodes, *Globodera rostochiensis* and *G. pallida*. *PANS*, 23, 178-189, 1977.

6. Kirjanova, E.S. and Krall, E.L., Plant parasitic nematodes and their control, Nauka Publishers, Leningrad, English Translation by American Publishing Co., New Delhi, 1980.

7. Caubel, G., Problems poses parie nematode des tiges (*Ditylenchus dipsaci*) on culture de mais, *Sciences Acad. Agric.* (France). 59, 627-640, 1973.

8. Meagher, J.W., Cereal cyst nematode (*Heterodera avenae*) : Studies on ecology and control in Victoria, Tech. Bull. 24, Dept. Agric., Victoria.

9. Anonymous, Estimated crop losses from plant parasitic nematodes in the United States, *Rept. Soc. Nem. Comm. on* Crop Losses, 1970.

10. Weischer, B. and Steudal, W., Nematode diseases of sugarbeet, In Economic Nematology (Ed. J.M. Webster), Academic Press, London, 49-65, 1972.

11. Suit, R.F. and Ducharme, E.P., Spreading decline of citrus, *Florida Dept. Agric. Bull.*, 1-20, 1967.

12. Cohn, E., Nematode diseases of citrus, In Economic Nematology (Ed. J.M. Webster), Academic Press, New York, 215-244, 1972.

13. Hollis, J.P. and Keoboonrueng, Nematode parasites of rice, In Plant and Insect Nematodes (Ed. W.R. Nickle), Marcel Dekker Inc., New York, 95-146, 1984.

14. Mamiya, Y., The pine wood nematode, In Plant and Insect Nematodes (Ed. W.R. Nickle), Marcel and Dekker Inc., New York, 589-626, 1984.

4

General Morphology of Nematodes

Plant parasitic nematodes as well as the free living soil and fresh water nematodes are microscopic, metamerically unsegmented (segmentation apparent in some cases is confined to cuticle only), thread like slender animals which are generally bilaterally symmetrical (condition whereby any two parts of the body are exact counterparts of one another) and vary in length between 0.2 to 10.0 mm. The majority of them, however, are less than 0.2 mm in length. Typically, the nematode body is cylindrical (elongate), sometimes fusiform; gradually tapering at the both extremities. The adult females of some groups exhibit a further modification in shape in as much as some may be pear, lemon, reniform or saccate shaped. For example, the females of the genera *Meloidogyne*, *Heterodera*, *Nacobbus*, *Tylenchulus* and *Rotylenchulus* assume one of these different shapes (Fig. 1).

The nematode body is circular in cross section (Fig. 2D), with the principal axis as the longitudinal axis. The width of the nematode body varies with the total length, the long thin forms being generally filiform, with almost the same body width throughout the length, *e.g.*, members of the genus *Ecphydaphora*. In the shorter nematodes, body width is narrow at the extremities and wide in the middle and may be as much as 15-16 per cent of the total length, *e.g.*, members of the family Criconematidae.

The nematode body mainly consists of three tubes, the outer tube or the body wall and musculature, the inner tube or the digestive tract and the reproductive parts which are embedded in the false coelom of the nematode body. There is no definite division of

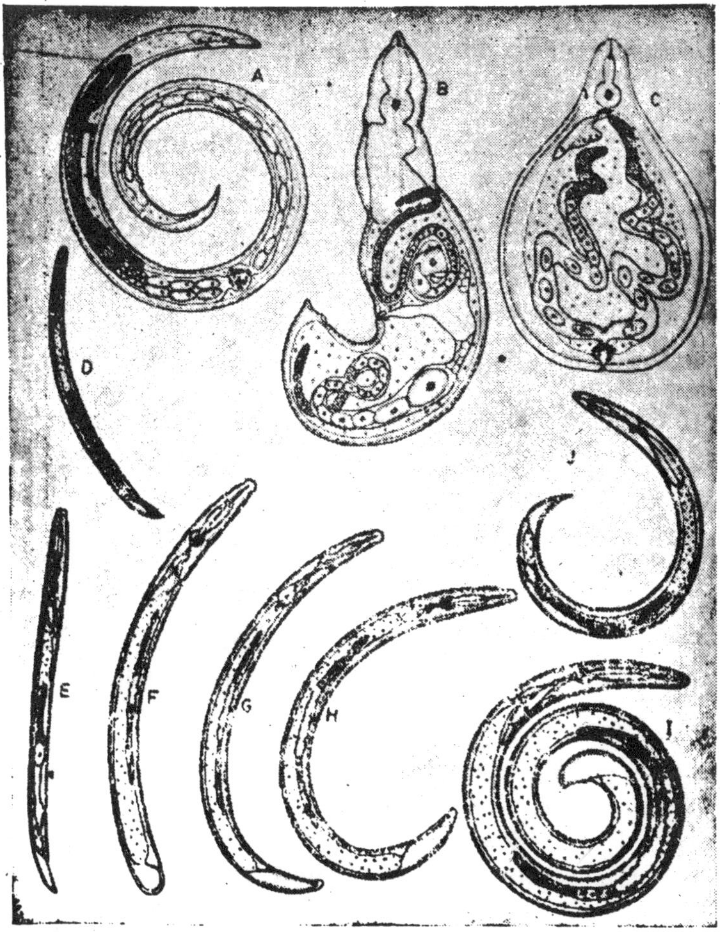

Fig. 1. Sexual dimorphism (A—D) and body postures (E—J)
A—*Anguina* ; B—*Rotylenchulus* ; C—*Meloidogyne* ;
D—Criconematid male ; E—*Pratylenchus* ;
F—*Hoplolaimus* ; G & H—*Tylenchorhynchus* ;
I—*Helicotolenchus* ; J—*Dorsalla* (After Jairajpuri, 1986).

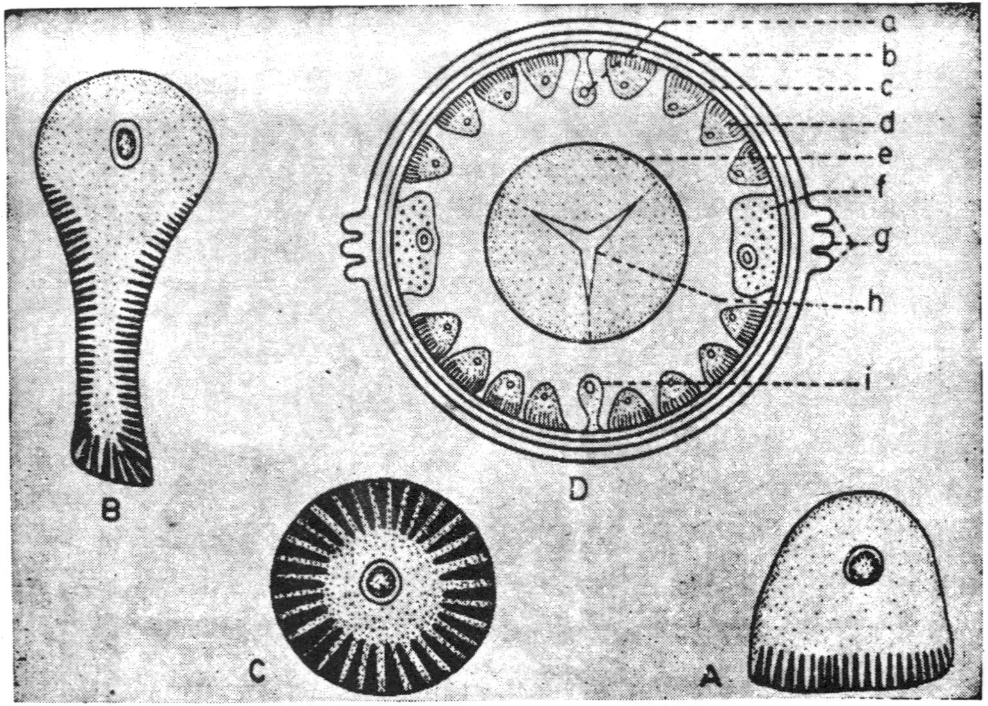

Fig. 2. Epidermis and somatic musculature.
A—Platymyarian; B—Coelomyarian; C—Circomyarian muscle cell; D—Cross section of body. (*a*) Dorsal chord; (*b*) Cuticle; (*c*) Epidermis; (*d*) Platymyarian muscle; (*e*) Oesophagus; (*f*) Lateral chord; (*g*) Lateral ridges; (*h*) Oesophageal lumen; (*i*) Ventral chord (After Jairajpuri, 1986).

parts of the nematode body, but for convenience sake, the body can be sub-divided into recognisable parts. The anterior end, bearing the mouth, pharynx and cephalic papillae, is termed as the head. This is followed by generally a well defined oesophagus, intestine and finally the tail which characterises the portion between the anus and tip of the nematode body. Longitudinally, the body is divided into four regions : dorsal, ventral and two lateral sides. When at rest, nematodes generally have a tendency to curve ventrally. However, the ventral side can be easily ascertained by locating the anal opening or the excretory opening or the vulval opening in adults. The position of the vulval opening is variable, usually located in the middle of the body but in certain cases it may be slightly above the anal opening.

Body Wall : The body wall is composed of cuticle, hypodermis and somatic musculature.

Cuticle (Fig. 3) : It is a protective covering and extends over the nematode body and into the various openings. It is non-cellular colourless and tough and may be smooth or deeply striated into annules as in the genus *Criconemoides* or marked by punctations or spines or various other markings. There may be longitudinal thickenings or expansion of the cuticle which are termed as alae. These may extend laterally throughout the length of the nematode body (termed as lateral field) or may be confined to the two extremes of the nematode body.

Normally, there are four moults during the life cycle of a nematode and with each moult, the old cuticle is shed off from the body as also the cuticle of the pharynx, vagina and rectum. In some instances, however, the old cuticle may be retained at the final moult as in the case of members of the genus *Hemicycliophora*.

External structures : *Lateral fields* : These help in the free movement of annules and are variable in number (two to six) in different genera and species. In a cross section, these appear as infolds of the cuticle. These are also referred to as incisures, alae, lateral striae, lateral alae, etc. Generally, these incisures start near about the position of the nerve ring and terminate near the anal region. The transverse striae are broken by these incisures but in some species, the transverse striae, at places, run across the incisures

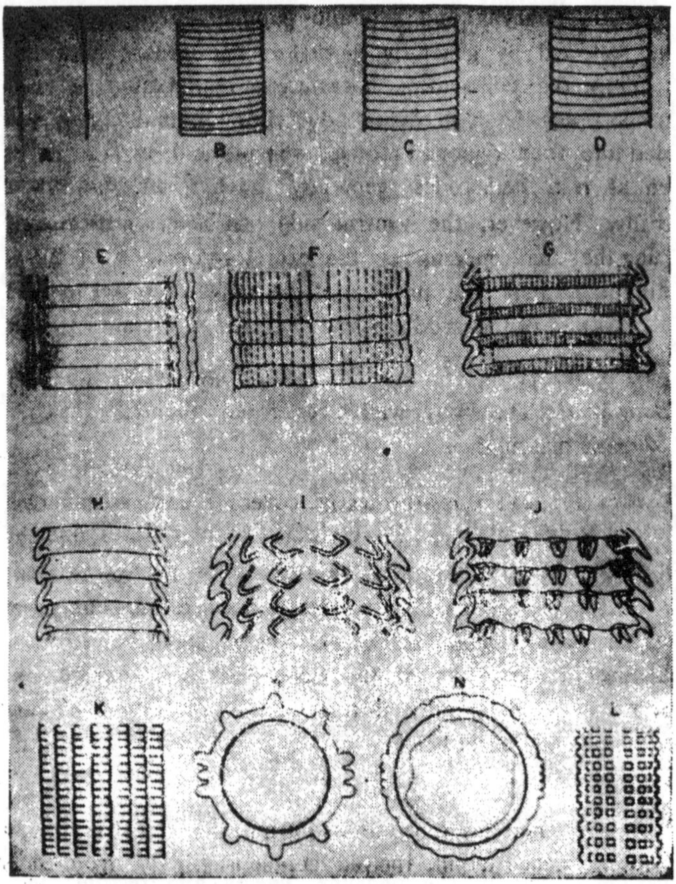

Fig. 3. Lateral fields.
A—*Dusulcius*; B—*Divittus*; C—*Tylenchorhynchus*;
D—*Quinisulcius*; E—*Merlinius*; F—*Aphelenchus*;
G—Irregular or incomplete aerolations;
H—Regular aerolations; I—*Hemicycliophora*
(After Jairajpuri, 1986).

giving rise to aerolation. In some genera of plant parasitic nematodes, the striae tend to disappear or are difficult to observe because of the huge gonads or obesity of the female body as in the genera *Anguina*, *Ditylenchus*, *Heterodera*, etc. However, in the larval forms, it can be observed easily. The incisures, which are often diagnostic in value, can be seen best in freshly killed nematodes. The most reliable method of ascertaining the number of lateral incisures is to cut a cross section of the nematode body near about the mid region and mounting it for an end-on view.

Cuticle layer. Histological studies have demonstrated that cuticle is a complex structure, which may be different with different groups of nematodes or even different in different growth stages of a single nematode species. Although, in the cuticle of *Ascaris lumbricoides*, as many as nine layers have been identified but in general, the external cuticle layers could be broadly grouped in three main layers : (1) cortical ; (2) median (matrix) ; and (3) basal (oblique fiber) layers. Recently, *Deontostoma* (Enoplia) has been proposed to possess the model cuticle since the genus is considered to possess nearly all the complements of strata[1]. According to this model, the nematode cuticle is divided into four fundamental layers—epicuticle, exocuticle, mesocuticle and endocuticle. Besides these four basic layers, anteriorly and posteriorly, some nematodes may have additional deposits called 'infracuticle'.

Chemically, the cuticle is apparently composed of, atleast in the case of *Ascaris lumbricoides*, albumins, a glycoprotein, a fibriod, a collagen and a keratin. Chemically, the external cortical layer shows characteristics of keratin because of its sulfur content and ratio of histidine, lysine and arginine. But x-ray defractor patterns shows its similarity with collagen. Quinones, polyphenols and polyphenol oxidase activity has been detected in some nematodes of *Heterodera* species. In a study on *Ditylenchus dipsaci*, it has been demonstrated that the imperviousness of the cuticle is due to a membrane which is thermolabile—probably a wax or sterol which disappears when the nematodes are killed by heat[2]. Under this layer lies the cortex followed by matrix which is rich in matricin (rich in sulphur). The fibre layer consists of collagan.

Fifteen amino acids in the cuticle component of *Meloidogyne javanica* and *M. hapla* alongwith presence of phenolic compounds and enzyme polyphenol oxidase have been recorded[3]. Qualitatively and/or quantitatively, the cuticle components may vary with different genera. This is reflected by their differential permeability to fixatives and stains. An electron microscopic study of the cuticle of *M. javanica* has shown that the cuticle consists of an osmophilic external cortical layer and an internal cortical layer which is morphologically different from the thick fiber layer[4]. The outermost layer is about 7 mμ thick and corresponds with a triple layered plasma membrane. The internal cortical layer is biochemically active and has enzymes and RNA. Generally, marine nematodes have greater permeability of the cuticle than the animal or plant parasitic nematodes. The cuticle of the living nematode may act as a semipermeable membrane for selective entry of metabolites.

As has been stated already, the cuticle may be smooth or marked by transversely striated lines to resemble segmentation. In *Pratylenchus* and *Tylenchorhynchus* species, the striae are fine in comparison to deep striations in *Hemicycliophora* and *Hoplolaimus* species. In Criconematinae, the striations become even more prominent and resemble almost segmentations. As a result of swelling of the females in the Heteroderids, the transverse striae are apparently modified into perineal patterns and zig-zag pattern.

Other cuticular markings are punctations suggested by some of the openings of pore canals[5] in the cortical layer and may occur either in transverse or longitudinal rows. The cuticle of *Heterodera* species is abnormally thick and contain pore canals[5] and polyphenol oxidase[6]. Spines, bristles, warts or setae are also cuticular markings and these may be localised or occur all over the body. Cuticle may be also inflated in some cases appearing as swellings in a blister like manner.

Besides transverse striations, longitudinal markings may be present on cuticle and these may be simply interrupting the transverse striations or may be ridges or alae. The term alae is used here to designate thickenings or projections extending throughout the nematode body or may be confined to the anterior or posterior regions. The longitudinal alae, usually four in number are sub-lateral in position. The lateral fields are two such longitudinal cuticular thickenings.

Besides the longitudinal alae, modifications may be present at the two extreme ends, the anterior region being known as cervical alae and mainly observed in animal parasitic nematodes. The caudal alae, which is located in the posterior region of the males, serves as the clasping organ at the time of copulation. However, it is not an essential feature of the males. It may be present in some genera and absent in others. It may also have genital papillae, in some cases, which serve as tactile organs. There may be six or more pairs as in Rhabditida or three or fewer as in Tylenchida.

Besides forming the top layer of the nematode body, the cuticle is involved also in the formation of lips, setae and sensory organs like deirids, amphids and phasmids.

The internal cuticle, which invaginates mouth, rectum, cloaca, vagina and excretory pore, is characterised by having epi and exo-cuticle.

Hypodermis. This is a thin layer beneath the cuticle and is a living cellular or syncytial layer. It secretes and maintains the cuticle and is joined by fibres to the somatic muscle cells. Typically, the hypodermis protrudes into the body cavity (pseudocoelom) at four positions, two laterally and one each at the dorsal and ventral positions. These are referred as lateral, dorsal and ventral chords respectively. The nuclei are located in these chords. The cells, intervening the two chords, lack nuclei. In primitive forms, hypodermis is cellular and is composed of relatively few cells. In the higher forms, the structure is more complex. Large amounts of fat and glycogen are stored in the hypodermis and it is thought that the hypodermis is one of the main reservoirs of glycogen in nematode body. The longitudinal body nerves and in some cases the lateral excretory canals are contained in the chords. In some nematodes, some of the hypodermal cells are modified to gland cells which are also associated with sensory structures like amphids and phasmids. In several free living forms, particularly in the members of Enoplida, Monohysterida, Chromodorida, unicellular glands (caudal glands) are present in the posterior part of the tail. These are three in number and have ducts which join together at the posterior end, opening externally. They release a secretion, which in contact with water, hardens to a thread like from enabling the nematode to cement itself to the substratum.

Pseudocoel. Underneath the hypodermis is the nematode body cavity called as pseudocoel because unlike the coelomic cavity, it is not lined with mesentry. The cavity is filled with a fluid (pseudocoelomic fluid) fibrous tissue and large fixed cells called pseudocoelomycetes.

Muscles

Somatic musculature (Fig. 2, A-C). Beneath the hypodermis, a single layer of cells, more or less spindle shaped, longitudinally oriented, and attached to the hypodermis throughout its length, comprise the somatic musculature. It is divided by longitudinal chords into longitudinal fields. The cells are arranged in rows of two, four or more and are variable in number. Earlier, the number of cells was considered to be of some taxonomic value but because of the fact these are quite variable in number, these are not considered to be of any taxonomic value. Based on the number of muscles cells in each quadrant, the nematodes have been classified mainly in two categories, meromyarian and polymyarian. Meromyarian is applied to the condition wherein only two to four longitudinal rows of muscles are present between the chords as against the other category (polymyarian) where large number of muscle rows (six or more) are present in each field. A third category, not used now, exists in earlier literature under the name holomyarian which incorporated those nematodes where the muscle layer was divided into two fields or was continuous because of presence of only two chords or no chords at all.

The musculature helps in the movement of the nematode by contracting and relaxing in the anterior as well as the posterior planes. The turgor pressure of the body contents and the elasticity of the cuticle oppose the contraction of the muscles. Waves of such contraction in the opposite plane, in the dorsal and ventral musculature, result in the typical serpentine movement of the nematode body.

The muscle cells are spindle or rhomboid shaped and consist of two zones, sacroplasmic and fibrillar. The former contains the nucleus and fibrillar network and in the latter, ribbon like fibres, separated from each other by sacroplasm, are present. Two types of muscle cells are recognised. When the ribbon-like fibres are present only on the side of the muscle cell, next to the hypodermis, it is referred to as platymyarian muscle cells. When the fibres extend to the sides of the muscle cells, partly enclosing the sacroplasm, then these are called coelomyarian muscle cells. However, various inter grades may be encountered between these two types of muscle cells but generally, nematodes having meromyarian muscle system have cells of platymyarian type and the polymyarian types have coelomyarian muscle cells.

Specialised muscles. In addition to the general body musculature, whereby motility and progression are affected, there are various special muscles such as those connected with anus, vulva, retractile muscles of the spicules and specialised muscles present in the male tail, which facilitate copulation.

EXCRETORY SYSTEM (Fig. 4 A, B)

The excretory system presents a varied picture in the phylum as a whole. It is simplest in the class Adenophorea where there is a single ventral excretory cell called rennete, which opens through an excretory pore on the mid-ventral line in the region of the oesophagus, by way of a short to long duct. In the class Secernentea, there are two lateral excretory canals, embedded in the lateral chords of the hypodermis. throughout most of the body length. They are connected, anteriorly and ventrally, by a transverse canal, forming a H or U shape. A very short duct connects the transverse duct with the excretory pore. The terminal excretory duct is cuticularly lined and can be observed in routine microscopic examinations. In the Adenophorea, the terminal excretory duct is not lined with cuticle (except in some Plectinae) thus making it difficult to see. In a few genera, including *Dorylaimus*, no excretory system has been recorded.

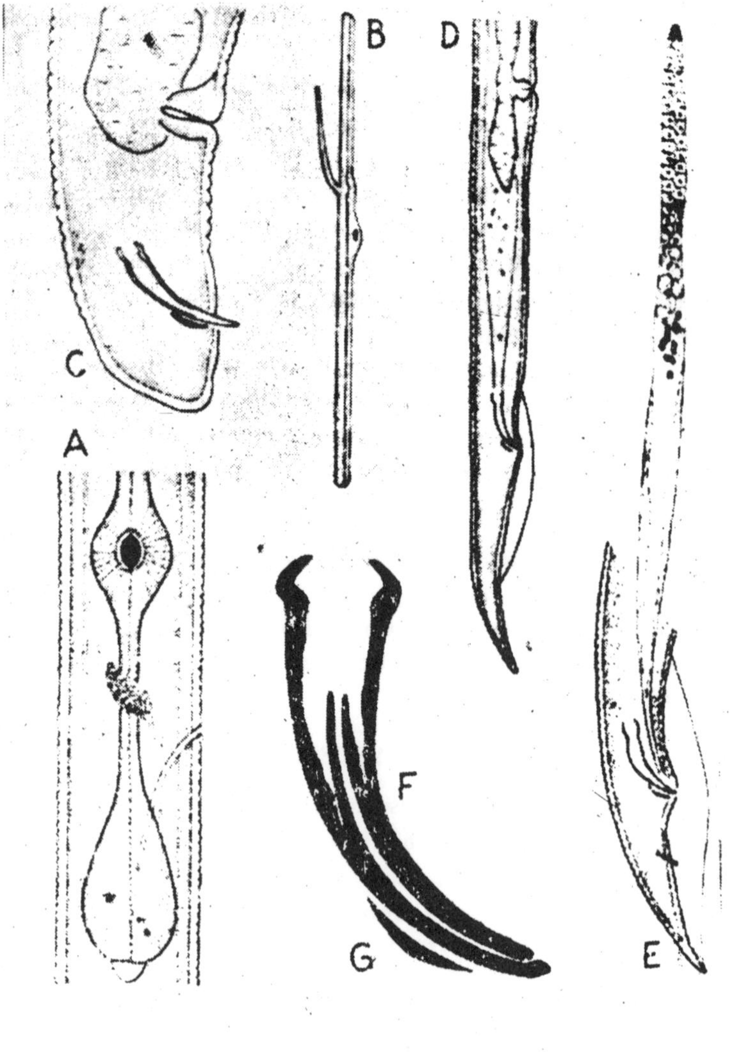

Fig. 4. Male reproductive organs, intersex, excretory system, etc. A—Oesophageal region showing nerve ring, excretory pore and duct; B—Tylenchid excretory system; C—Intersex (*Meloidogyne*); D—Intersex (*Ditylenchus*); E—Male reproductive organs; F—Spicule; G—Gubernaculum (After Jairajpuri, 1986).

Considerable excretion through the digestive tract may occur in all nematodes.

The separation of the two classes of the Phylum into Adenophorea and Secernentea (Aphasmidia and Phasmidia) is based, to some extent, on the type of the excretory system present.

DIGESTIVE SYSTEM

The digestive tract consists of the following principal regions: stoma or buccal cavity, oesophagus, intestine, rectum and anus. The oral opening of the stoma is surrounded by lips, usually six in number, designated as subdorsal, lateral and subventral. In plant parasitic nematodes, the number of lips as well as the sensory structures may be reduced in numbers. The stoma may be very simple or quite elaborate—a simple plain walled tube with a spear or complicated cavity with teeth or denticles, etc., for puncturing, masticating or tearing. The anterior part of the digestive tract, comprising of buccal cavity (stoma), oesophagus and the oesophagus-intestine valve, is also referred as the foregut or stomodeum or stomatodaeum. The mid-gut or the intestine is the mesentron. The hindgut or the proctodaeum comprises of the intestine-rectal valve, the rectum in the females and cloaca and its associated structures in the case of males. The oesophagus and the rectum are differentiated from the intestine by their cuticular lining. The digestive tract as such, from the oral to anal openings, appears as a simple tube. The characters and the morphology of the stoma as well as the oesophagus are of taxonomic value and have a primary importance in all systems of classification.

Stoma or the buccal cavity. The oral opening, which is terminal, leads directly to the stoma. In the different nematode groups, it may vary in shape, size and morphological details. In most of the nematodes, the stoma consists of two parts, the anterior section referred to as cheilostom and the posterior one called as oesophastom. According to a nomenclature[7], based on the genus *Rhabditis*, there are five parts of the stoma, with cheilostom as the anterior most part. According to this system, the oesophastom is divided into protostom (prorhabdion) and telostom (telorhabdion).

The protostom is further subdivided into three parts, protostom (prorhabdion), mesostom (mesorhabdion) and metastom (metarhabdion). Parts of the stoma may be sclerotised and the muscular linings, in each case, are designated as indicated in the parentheses above for the different sections.

All plant parasitic nematodes possess a stylet. In the Secernentea (Phasmidia), this is considered to be formed by the coming together of the sclerotisation of the buccal capsule and accordingly termed as stomatostylet or stomatostyle (Fig. 5). In Adenophorea (Aphasmidia), the anterior part of the stylet is representing an enlarged tooth that originates in the oesophagus wall. This is termed as odontostylet or odontostyle. The posterior part of the stylet is derived from the lining of the oesophagus and is termed as odontostylet extension. After initiation of the odontostylet from the oesophagus wall, it moves forward to its position in the stoma with each successive moult.

Typically, the stylet of a plant parasitic nematode is composed of three main parts—an anterior conical part, the middle shaft and the three knobs, one dorsal and two sub-ventral. The protractor muscles for the stylet are attached to the posterior part of the stylet. At each moult, during the larval stages, the anterior part of the stylet is shed off alongwith the cuticle and the remaining two parts disappear. The new stylet is formed with the emergence of the conical part first followed by the latter two parts. In the plant parasitic nematodes, stylet is an essential part of the feeding process by the nematodes. The protractor muscles, attached to the stylet, help the stylet in movement, in and out of the stoma. In some of the plant parasitic nematodes, stylet is weak with no basal knobs (genus *Aphelenchus*) or with weak knobs (as in genus *Aphelenchoides*) or with strong massive knobs (as in genus *Hoplolaimus*). The males of some groups have a degenerated stylet or do not possess any stylet but the juvenile forms always exhibit a stylet. In the tylenchs, the stylet is a hollow, tripartite and cuticular structure while in the dorylaims the stylet is generally not hollow.

Oesophagus. Next to stoma is the oesophagus which varies in structure in different groups of nematodes. It is, therefore, of great

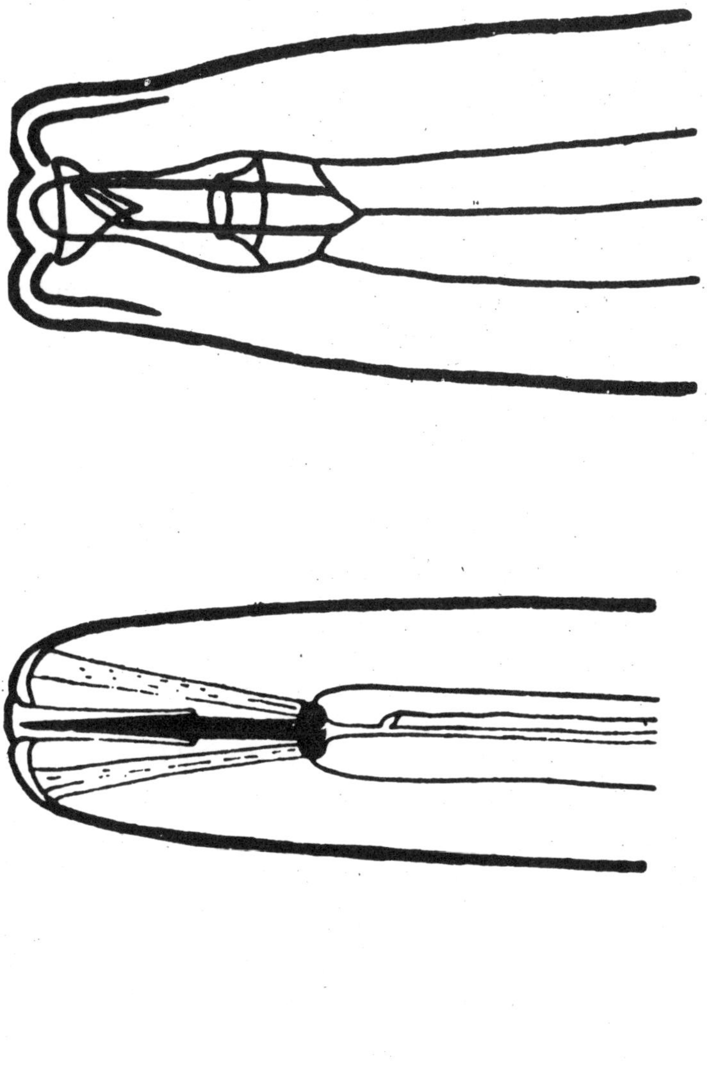

Fig. 5. Stylet forms. Left—Stomatostylet; Right—Odontostylet.

taxonomic value. It is a tube lined by thin cuticle and covered by a membrane. The lumen of the oesophagus is triradiate and is extended into three symmetrically arranged longitudinal grooves that partially divide the oesophagus wall into three sections, one dorsal and two ventro-lateral (sub-lateral). However, the triradiate oesophageal lumen may not be present throughout the tube in all the parasitic nematodes. It is cylindrical in the anterior oesophageal part of *Xiphinema index* and from the base of the stylet knobs to the valve of the metacorpus in *Ditylenchus dipsaci* and *Macroposthonia xenoplax*.

The oesophagus has one or more muscular swellings known as bulbs. These may be located near the mid length (designated as median bulb) or may be at the end of oesophagus (designated as posterior or cardiac or end bulb). According to the shape and position of the bulb, the oesophagi of the nematodes are divided into different categories (Fig. 6) : (1) Cylindrical--It is the most primitive muscular type and is encountered in marine nematodes of the family Enoplidae (for example *Enoplus*) or as in Mononchidae (*Mononchus*). From this basic type, other modifications in the structure of the oesophagi, possessing one or two bulbs, may be derived. (2) Dorylaimoid type—Anteriorly slender and non-muscular with a swollen muscular glandular posterior part. This type is encountered in the plant parasitic genera of *Longidorus*, *Xiphinema*, *Trichodorus*, etc. (3) Bulboid type—Slender anterior part ending with a posterior end bulb as in *Ethmolaimus*. (4) Rhabditiod type—Anteriorly, it has a wide region (corpus), usually leading to a median pseudobulb, followed by a narrow region (isthmus) and succeeded by an end bulb with valvular apparatus (as in members of the genus *Rhabditis*). (5) Diplogasteroid type—The anterior muscular region terminates in a median bulb. It is succeeded by a posterior glandular region forming a distinct bulb without a valvular apparatus (as in *Diplogaster*). (6) Tylenchoid type—Most of the plant parasitic nematodes belong to this type. It has a narrow oesophageal tube attached to the base of the stylet and is enclosed in a large thin walled tube. In most of the genera, there is a muscular median bulb with an ovoid valve (this bulb is much reduced or even absent in members of Neotylenchidae). The posterior part of the oesophagus is glandular and the glands may form a distinct buccal bulb or a lobe overlapping the anterior part of the intestine. The duct of the dorsal oesophageal gland empties in the oesophageal tube wall, anterior to the median bulb, usually very near the base of the stylet. (7) Aphelenchoid type—It is similar to the tylenchoid type except for the fact that the dorsal oesophageal gland empties in the lumen of the oesophagus in the median bulb.

The oesophagus may be connected with the intestine with a structure termed as oesophago-intestinal valve which functions as an organ to impede the flow back of food from intestine to the oesophagus. This valve is cuticularly lined and may be very small or very large as well as conspicuous. It has been termed also as cardia. In some cases, as in *Plectus*, this valve may go deep in the intestinal lumen and may be wholly surrounded by the intestinal wall.

Intestine. It is composed of a single layer of epithelial cells and is usually a straight tube in contrast to the reproductive organs which may be reflexed or coiled. The number of the intestinal cells are extremely variable from relatively few in number to several thousand. In Tylenchida, the number is small, ranging from 16 to 24, while the Rhabditids have from 18 to 64. On the basis of number of intestinal cells, two nematode categories have been defined. When the number of cells are 128 or less, the nematodes are categorised as oligocytous and when it is more than 128, the term polycytous is used. In oligocytous types, represented by plant parasitic group, the cells are large and multinucleate. They contain stored food materials like fats, glycogen, fatty acids and proteins, in addition to the protoplasm and nuclei.

Posteriorly, the intestine leads into the rectum which is lined by an invagination of body cuticle and opens at the anus. The rectal glands, three in number, open into the rectum, one dorsally and two subventrally. In male nematodes, the reproductive and the digestive systems join posteriorly in the ventral wall of the rectum to form the cloaca. The rectum is thus, wholly or in part, a cloaca in male nematodes. In both sexes, the rectum empties posteriorly through the anus aperture on the ventral surface of the body.

NERVOUS SYSTEM

The nervous system in nematodes comprises essentially of three parts namely : (1) the central nervous system, (2) the peripheral nervous system, and (3) the peripheral nerve net.

The Central nervous system. The most recognisable part is the nerve ring which encircles the isthmus region of the oesophagus. The nerve ring is associated with a number of ganglionic masses, with the ventral ganglion as the most conspicuous. There are six nerve trunks extending anteriorly from the nerve ring and innervating the sensory organs, tracto-receptors (sensory papillae) or setae and/or chemoreceptors (amphids), at the anterior extremity. The nerves running posteriorly are connected with each other by means of commisures.

The paired post-anal lateral sensory organs (phasmids), the ventral supplementary organs of males and the caudal papillae are all innervated by branches from one or the other of the main nerves. The nerves themselves and their branches can not be seen without special methods for demonstrating them or by means of sectioning.

Peripheral nervous system. These innervate deirids and hemizonid in addition to amphids and genital papilla. In routine microscopic examinations, the only part of the nervous system visible is the nerve ring whieh surrounds the isthmus region of the oesophagus. Even this is often difficult to observe.

Periheral nerve net. In Adenophorea, in addition to the central and peripheral nervous systems, a peripheral nerve net also exists which links somatic and papillae, which, in turn, are connected to the sensory neurons of the peripheral nervous system.

REPRODUCTIVE SYSTEM

Nematodes are generally amphigonus (dioecious), existing as separate males and females and are oviparus. The sexes can be easily identified by the primary and secondary sexual characters. The males possess the copulatory apparatus with prominent spicules at the posterior end of the nematode body while the females possess the vaginal aperture, on the ventral side, in the middle or the posterior half of the body. Rarely, it may be located in the anterior region of the body. Other distinguishing features of males, which may or may not be present in all the cases, are comparatively smaller size, curvature of the posterior end and presence of a caudal alae or bursa and genital papillae and other accessory copulatory structures. In the marine and most of the plant parasitic nematodes, both the sexes occur in almost equal proportion. In the terrestial and fresh water forms, particularly the latter, the females predominate in the population. The scarcity or absence of males indicates a tendency towards hermaphroditism or parthenogenesis. The hermaphroditism is usually of the protoandric type—the gonad releases the sperms first which are stored in a special structure and subsequently when an ovum is produced by the same gonad, the stored sperms fertilise the ovum to form an egg.

Apart from the sexual differences, the males and females, in some genera of plant parasitic nematodes, exhibit extreme sexual dimorphism. For example, in the genera *Heterodera* and *Meloidogyne*, the females are swollen to lemon or pear shape and become a reproductive sac and the males remain slender and worm like. In other forms, both the sexes may remain vermiform but the males may lack a distinct feeding apparatus as in the Criconematids or the genus *Radopholus*. Intersexes are also encountered in some populations like *Ditylenchus triformis* and *Meloidogyne* species. In such cases, the females do not develop fully and possess rudiments of male characters. Probably, this results from partial sex reversal under environmental influence or nutritional stress.

Gonads. It is essentially similar in both sexes of all nematodes. It is composed of one or two tubular gonads, varying greatly in length. These may be also straight or sinous, reflexed or coiled back and forth. Females with two complete genital tubes are referred to as didelphic (diovarial) and those with only one tube are called monodelphic (monovarial). In the didelphic forms, if the uteri are opposed at the point of origin, then they are called amphidelphic. If they are parallel and anteriorly directed at the origin, then they are called prodelphic. If the uteri are parallel and directed posteriorly at the point of origin, then the condition is referred to as opisthodelphic (Fig. 7).

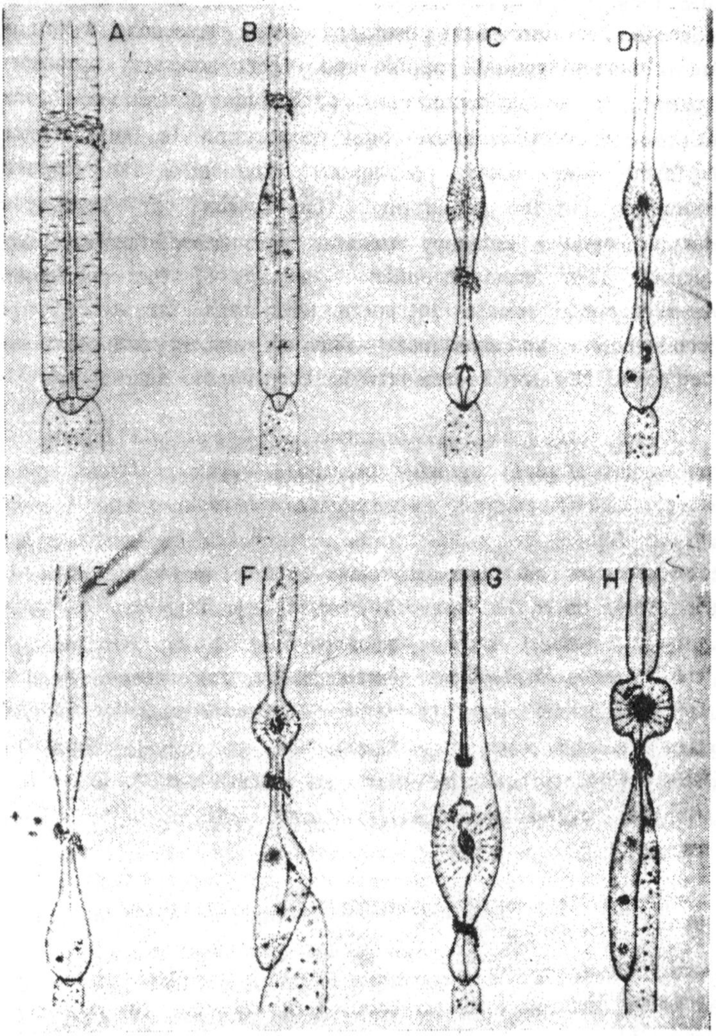

Fig. 6. Different types of oesophagi.

A—Mononchid ; B—Dorylaimoid ; C—Rhabditoid ; D—Tylenchoid ; E—Neotylenchoid ;
F—Hoplolaimid ; G—Criconematoid ;
H—Aphelenchoid (After Jairajpuri, 1986).

Male Reproductive System (Fig. 4 E, F). The male nematodes may have one testis (monorchic) or two testis (diorchic). Usually, there is a single testis in the male. However, when two testes occur, these are oppositely oriented, except in the genus *Meloidogyne* where they are parallel. When two testes are present, one usually extends forward and the other in the backward direction in the body cavity, but they join medially in the vas deferens, which runs posteriorly ventral to the intestine and finally narrows down to an ejaculatory duct which opens in the rectum to form a cloaca. The main part of the male reproductive duct is the vas deferens which is usually composed of tubular and glandular regions. The testis can be divided into germinal and growth zones. The maturation of the sperms takes place at the end of the growth zone. The sperms may be amoeboid, rounded, flagellate or conical. With fewer exceptions (for example in the genus *Trichinella*), the males are provided with-copulatory spicules. Often the spicules are accompanied by an accessory piece, the gubernaculum, which is formed by the sclerotisa tion of the dorsal wall of the spicule pouch.

Female Reproductive System (Fig. 7). The main parts of the female reproductive system are ovaries, which are tubular sac consisting of a germinal zone where rapid cell divisions take place ; the growth zone where the oogonia gradually increases in size ; the oviduct which follows the ovary and is a constricted thick walled region. The uterus is a broad tube. The eggs are stored in the uterus. The beginnings of the uterus usually function as a spermatheca or seminal receptacle. In some nematodes, a slight enlargement in the proximal part of the oviduct serves as a spermatheca or seminal recptacle. The uterus is covered also by a muscle layer. The eggs, after fertilization and maturation, and to some extent embryonic development also, pass out through the vaginal opening, the vulva. The nematodes, with a single ovary, sometimes possess also the rudiments of the second genital tube which may function as a spermatheca. The rudimentary tube is known as a post-vulval uterine sac.

Reproduction. Males are important for reproduction in the bisexual species. In such cases, the males are as abundant as the females in the populations. In some species, the females can reproduce parthenogenetically, as in genus *Meloidogyne*. In these cases,

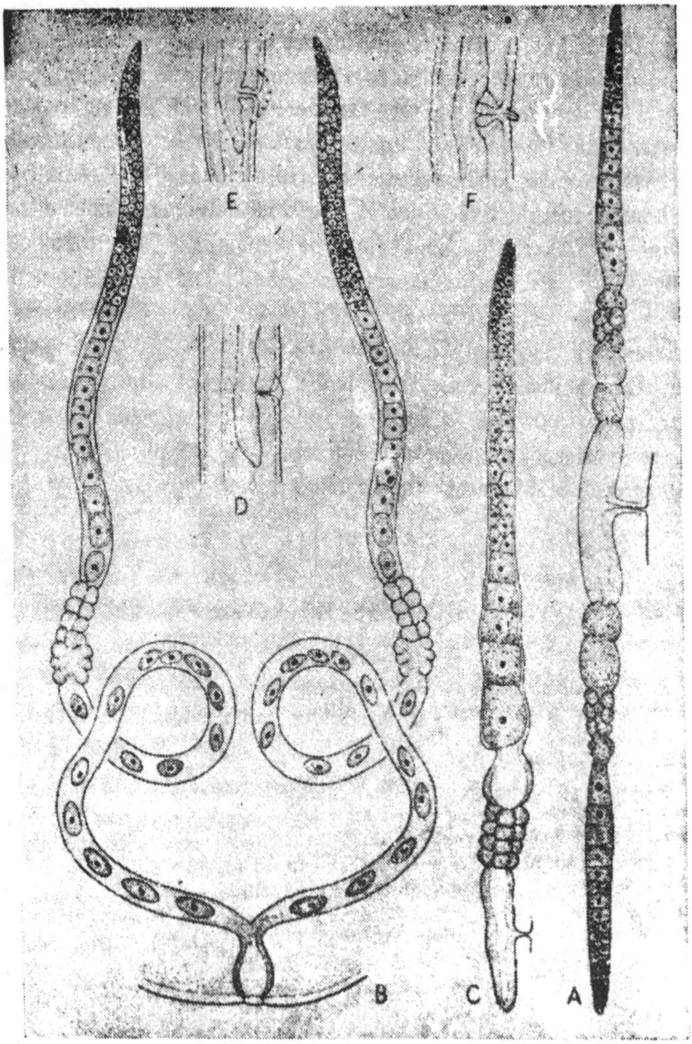

Fig. 7. Female reproductive organs.
A—Didelphic, amphidelphic ; B—Didelphic, prodelphic ; C—Monodelphic, prodelphic ; D & E—Vulval flaps ; F—Epiptygma (After Jairajpuri, 1986).

the nematode populations are either devoid of males or if present, their number is small or the males are degenerated type. The absence of males in the population or any of the above possible conditions, indicates a tendency towards hermaphroditism or parthenogenesis, In the latter case, there is self-fertilization with the sperm, the eggs being produced in the same gonad (syngonic type of reproduction). In an another type, the egg production by the gonad is preceded by the production of sperms by the same gonad (protandry type of reproduction). Males, if present, in such cases are apparently functionless.

Aberrent sex forms may be found in some populations. Fully developed females may have rudimentary male characters and *vice versa*. Such individuals are termed as intersex. Female intersexes, in some cases, have been recorded to have copulated with males and have produced eggs. Intersexes seem to result from partial sex reversal under environmental influences (Fig. 4 C, D).

Sex Determination. The most common mechanism of sex determination in nematodes is that of XO-XX type. That males, in this system, have atleast one chromosome less than the females. Epigenetic factors may also control production of male.[1] It is also possible that because of over crowding in the populations or lack of available nutrients there may be more males produced than females.

Eggs. Nematode eggs are generally ovoid but may vary in shape. The eggs have three membranous layers : (1) the external protein layer secreted from the uterine wall ; (2) a chitinous membrane secreted from the egg itself ; and (3) a vitelline membrane. Egg surface may be smooth (as in most of the plant parasitic nematodes) or may bear various specialised structures (as in some animal nematodes).

Embryonic Development. After release of eggs from the female body, a series of divisions take place. Initially, it starts with a transverse division into two equal cells followed by other quick divisions till a multicellular stage is reached. Soon, the first stage larva develops inside the egg. Generally, the first moult occurs inside the egg resulting in the development of second-stage larva. It remains coiled inside the egg till appropriate time for emergence or hatching. The second-stage larva, at this time, resembles the adults except for size and lack of reproductive parts. Before emergence from the eggs, the second-stage larva moves vigorously within the egg shell, probing with its stylet, till it finds a weak spot on the egg surface and then emerges after puncturing it.

Post-embryonic Development. The various developmental stages of the larvae are intervened by moults in which the entire cuticle including the lining of stoma, oesophagus, vagina, rectum and cloaca are shed off. There are generally three moults after the emergence of the second-stage larva and before the adult stage. In most of the plant parasitic nematodes, the genital primordium can be located in the second-stage larva and sex differentiation can be observed in the late second stage or the third stage larva.

CIRCULATORY AND RESPIRATORY SYSTEMS

These systems are not known to be present in the nematodes. The movement of the body cavity fluids, apparently, serve these purposes.

REFERENCES

1. Maggenti, A.R., General Nematology, Springer-Verleg, New York, 372 pp., 1981.

2. Chitwood, B.G. and Chitwood, M.B., An Introduction to Nematology, Monumental Printing Co., USA, 213 pp., 1950.

3. Bird, A., The adult female cuticle and egg-sac of the genus *Meloidogyne* Goeldi, Nematologica 3, 205-212, 1958.

4. Bird, A. and Rogers, G.E., Ultrastructural and histochemical studies of cells producing the gelatine matrix in *Meloidogyne*, 11, 231-238, 1965.

5. Bird, A.C., The structure of nematodes, Acad. Press, Lond. 318 pp., 1971.

6. Ellenby, C., Nature of cyst wall of potato root eelworm, *Heterodera rostochiensis* and its permeability to water, Nature (Lond.), 157, 302-303, 1946.

7. Steiner, G., The nematode, *Cylindrogaster* and its relationship, J. Parasitol, 20, 68, 1933.

5

Life History

The stages in the life cycle of plant parasitic nematodes start with the egg, the embroynic as well as post-embryonic stages generally following more or less a similar pattern in all the cases. After a series of transverse and longitudinal cell divisions, the first-stage larva, lacking stylet is formed. However, in the dorylaims (species of *Trichodorus*, *Xiphinema*, *Longidorus*, etc.), a week stylet may de discernible. In majority of the plant parasitic nematodes, except the dorylaim group, the first moult occurs within the egg shell and it is only the second-stage larva which normally emerges as the infective stage. In the dorylaim nematodes, the first-stage larva itself emerges from egg followed by a quick moult into second-stage larva.

Before breaking through the egg shell, the second-stage larva moves within the egg shell and probe the wall surface with the help of its stylet, for locating a weak spot -for continued stylet thrusts for puncturing the wall. In other cases, enzyme secretions like lipase, chitinase, etc., may help in the weakening and dissolution of multilayered egg shell to release the larva.

In the endoparasitic species of genera *Heterodera*, *Meloidogyne* etc., the second-stage larva enters the host roots and further life processes proceed inside the roots (Fig. 1). These larvae require feeding on host roots before proceeding with the other developmental stages. In the case of *Rotylenchulus reniformis*, however, the second-stage larva does not require feeding for further development. Therefore, further superimposed moults occur in the soil itself. In most of the cases, a certain period of feeding may be necessary before the second moult takes place. This feeding period may be a short or prolonged one, depending on the nematode species. In the case of *Anguina tritici*, the second-stage larva requires atleast 60 days feeding on growing point of wheat seedings before it is able to moult to third-stage larva, while in other cases it may vary from 3-10 days or more depending upon the environment and availability of host[1]. Generally, feeding is considered an essential pre-requisite to moulting. However, in some cases, the larvae, after successive moults, retain the previous cuticle (*e.g.*, in *Meloidogyne*) and when this occurs, the nematode is not able to feed and there is very little growth in size.

Normally, there are three moults outside the egg before the larva becomes adult. Thus, there are four moults, four larval stages and an adult stage. In several species, both male and female adult nematodes occur in almost equal numbers in the population. The role of the male nematodes is only for copulation with the females. In these bisexually reproducing species, sperm from the male is transferred to the female during copulation. Fusion of sperm and egg occurs within the female body resulting in the formation of an embryo. In some species, even though the males may be present in the populations, they have not been recorded participating in fertilization of females, which develop parthenogenetically. Males of the most of the plant parasitic nematodes are not known to feed on host root tissues.

The second-stage larvae, in most of the cases, are the infective stage which then initiate, the life processes. In the case of *Rotylenchulus reniformis*, however, the pre-parasitic young female nematode is the infective stage. The other nematode stages, in the case of ectoparasites and also in some endoparasites like *Pratylenchus* are also capable of causing infection.

FEEDING

Feeding by nematodes is a process involving attraction, orientation and contact to host root system, probing and thursting of stylet in root cells, ingestion of cell contents and retraction of stylet in the case of ectoparasite. The morphological structures of the nematodes, involved in the feeding processes, are the stylet, parts of the oesophagus, secretory sub-dorsal and ventral oesophageal glands as also valves at matacorpus and oesophageal-intestinal junction which regulate the flow of injested food as well as the salivary juices.

Nematodes differ in their specificity for hosts and for parts of the host they feed on. In number of cases it has been positively demonstrated that nematodes are attracted to host roots, presumably in response to some stimuli. However, there are cases also that nematodes may be attracted towards non-host roots or organisms associated with the roots. The parts of the roots most attractive to nematodes are the tip, zone of elongation and in some cases zone of root hairs. On reaching the roots, most nematodes tend to explore the surface, making occasional stylet probes till a suitable site is located. Once this is achieved, the stylet probes increase in frequency and vigour and correspondingly the nematode movement is slowed down. The nematode processes its lip region against the root surface, and probably after it receives some stimulus (most probably chemical), the stylet thursts increase in vigour for penetration and feeding. The type and duration of stylet thursts may vary with species. In Criconematids, a firm grip is maintained between lip of the nematode and host cell by mucopolysaccharide adhesive plug originating from nematodes. In some of the dorylaims, especially species of *Xiphinema* and *Longidorus*, the stylet is attenuated and the anterior part of the oesophagus is quite slender. To regulate the flow of the injested food material, two musculature system have been recorded in *X. index*, consisting of a sheath composed of muscles and basement membrane surrounding the oesophageal bulb[3]. This may be visualized as additional pressure plate, besides the pseudocoelomic pressure, to regulate flow of sap from host cells to nematode.

One of the important aspects in feeding is the breaking down of the host cell barrier before food can be injested. It has been demonstrated in large number of cases that the nematodes are

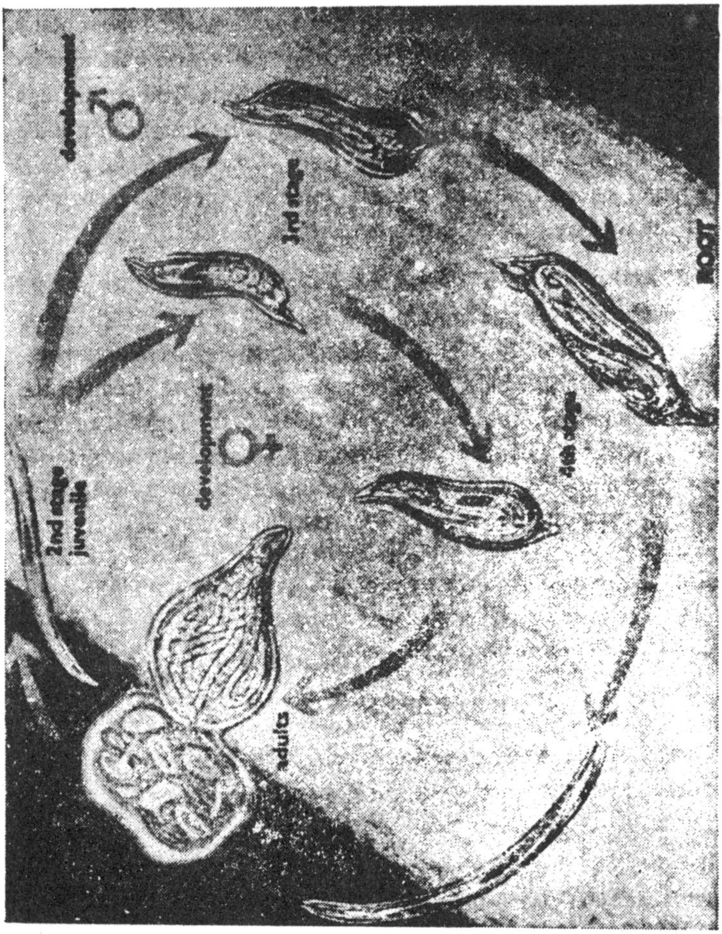

Fig. 1. Life processes of an endoparasitic nematode (*Meloidogyne* species).

equipped with enzymes capable of hydrolising components of plant cell wall. High level of cellulolytic activity has been recorded from several stylet bearing nematodes. The plant parasitic nematodes have a higher activity of the enzyme than the fungal or bacterial feeders. One of the hydrolytic enzymes, which has received considerable attention is pectinase and which has been reported from many plant parasitic nematodes. The successful feeding on host cells by *Ditylenchus dipsaci* depends largely on the nematode ability (through pectinase) to macerate host tissue through dissolution of middle lamellae without causing death to the cells. Since pectic compounds are important components of middle lamellae, pectolytic enzymes facilitate the feeding process by the nematode.

The secretions from the nematode glands are either surged forward, at the time of injection in host tissue, by body contraction movement (as in ectoparasites) or released slowly (as with endoparasites). The secretions help in digestion of food. At feeding, the forepart of the oesophageal lumen gets dilated for sucking of fluid juices followed by posterior dilation and simultaneous anterior closure. The nematode stylet which injects secretions in host tissues has also the capability of preventing viscous or granular materials from blocking the stylet. With some nematodes, the food being injested is predigested with the help of the secretions, while in other cases, the juice may be taken as such with digestive process taking place inside the lumen of the oesophagus.

REFERENCES

1. Gokte, N., Studies on biology and host-parasite relationship of the ear-cockle nematode, *Anguina tritici* on wheat, Ph. D Thesis, I.A.R.I., New Delhi, 1987.
2. Wright, K., The histology of the oesophageal region of *Xiphinema index* as seen with the electron microscope *Can. J. Zool.*, 43, 689-708, 1965.

6

General Symptoms

The plant parasitic nematodes are obligate parasites and except for a few species attacking buds, stems, leaves or other above ground plant parts, most of them feed on subterranean plant structures. Based on the mode of feeding, they have been classified as ecto or endo-parasites. Nematode-feeding on plant tissues may result either in mechanical or biochemical injuries, which are directly or indirectly responsible for symptom-manifestations of the disease. Mechanical injury occurs as a result of continuous thursting of stylet into the cells of the host plants while the biochemical type of injury is the resultant effect of the release of nematode salivary juices in the plant cell. The process of feeding is initiated by secretion of salivary juices into the plant cell (through lumen of the stylet), which apparently facilitates extra-corporeal food digestion necessary before it is injested into the nematode body. Apart from the assistance provided for food digestion, the salivary juices may bring about alterations in the plant parts. For example, cell division in the merismetic tissues may be stopped or slowed down or the cells surrounding the stylet may exhibit hypertrophy or hyperplasia or may die resulting in formation of lesions or necrotic tissues. Both the mechanical as well as the biochemical injuries manifest themselves in the form of specific morphological changes in the plant tissues. The symptoms of injuries, caused by plant parasitic nematodes can be divided, therefore, into two categories: (1) above ground; and (2) below ground.

Above Ground Symptoms. These can be either due to damaged root system, manifested in the form of weakened plant condition, or

Fig. 1. Crinkled and twisted leaves caused by *Anguina tritici* on wheat. 1—Healthy leaf; 2-5—Crinkled and twisted leaves.

direct damage to plant parts by nematodes. The symptoms, expressed as result of nematode feeding on roots, are usually non-specific and are due to inadequate or improper functioning of root systems.

The affected plants may exhibit leaf chlorosis, reduced vigour and in ability to tolerate adverse conditions (lack of moisture, temperature conditions, non-availability of proper nutrients), slow die-back, smalling of petioles and leaves, wilting under dry conditions, reduced tillering. One of these conditions or combinations of one or more factors may ultimately lead to reduction in quality as well as quantity of seed and fruit. Some nematodes feed on the aerial plant parts which may result in injury expressions in buds, growing points, flower primordia, leaves or stems.

1. *Devitalised Buds.* Infestations by *Aphelenchoides parietinus, A. besseyi, A. fragariae* may cause this type of injury. *A. parietinus* injury to strawberry plants or cotton seedlings may result in killing of buds or growing point stopping further growth of the affected tissues. The term 'blind' has been used for such conditions. *A. besseyi* on strawberry causes killing and dropping off of flower buds.

2. *Crinkled Stem and Foliage. Anguina tritici* larvae feed on the growing point of wheat seedlings without killing it. The developing leaves become twisted and crinkled (Fig. 1). *Aphelenchoides besseyi* on strawberry causes a condition known as 'crimp' 'summer crimp' or 'summer dwarf'. The affected plants show crinkled folded leaves, abnormally dark in colour and are quite brittle. Another species, *A. fragariae*, is also associated with strawberry disease known

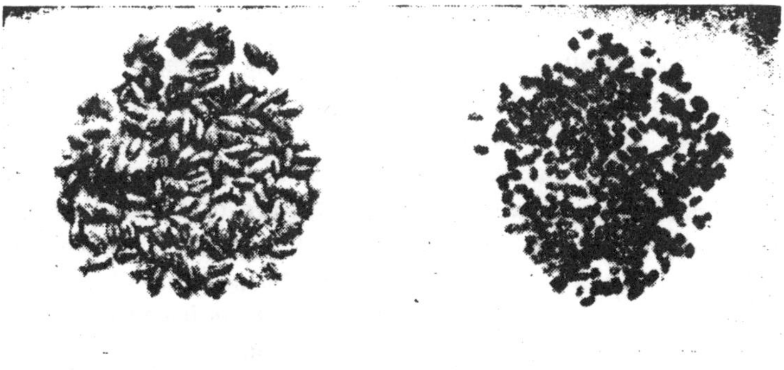

Fig. 2. Nematode seed gall—*Anguina tritici.*
Left—Healthy wheat grains ;
Right—Nematode galls.

as 'spring crimp'. Crinkled leaves can be associated also with infestations of *Nothotylenchus acris* on alfalfa and *Anguina tritici* on wheat.

Fig. 3. Chrysanthemum leaves showing lesions caused by *Aphelenchoides fragariae* or *A. ritezema-bosi*.

3. *Seed Galls.* The most outstanding example under this category is that of *A. tritici* on wheat. The nematode larvae enter the flower primordia which instead of normal development into seed, change to rounded galls (Fig. 2) which are green and soft in the initial stages but with crop maturity, turn to brown-black hard structures. The other examples are : *A. agrostis* on *Agrostis tenuis* and *A. stolonifera* ; species of *Festuca, Phleum, Poa.* Elongate galls are formed in the earhead. Galls formed on Festuca grass are recorded to be fetal to cattle, sheep, horses, etc.

4. *Leaf Galls.* Some species of *Anguina* produce galls on leaf surfaces as on *Erharta longifolia* by *A. australis* ; on *Balsamorrhiza*

sagittata by *A. balsamophila* ; on *Festuca ovina* and *F. rubra* by *A. graminis* ; on *Microlaena stipoides* by *A. microlaenae* ; on *Achillea millefolium* by *A. millofolii* ; on *Cynodon transvalensis* by *A. tumafaciens.*

5. *Leaf Spots and Lesions.* The foliar nematodes enter through leaf stomata causing destruction of leaf parenchyma which may become apparent in the form of leaf spots or lesions. *Aphelenchoides ritzema-bosi* and the fern and begonia races of *A. fragariae* cause leaf spots on chrysanthemum and begonia respectively (Fig. 3). The spots are first discernible on the underside of the leaf surfaces as small yellowish areas, later turning to brown colour and ultimately to black colour. The spots may spread and coalesce together till the entire leaf is destroyed.

6. *Necrosis and Discolouration of Stem and Foliage.* The narcissus race of *Ditylenchus dispaci* causes small, yellowish, slightly raised pimple like spots on leaves (narcissus) which are referred to as 'spikkels'.

Discolouration of stem and foliage may range from light or pale yellowishness to various shades of reddishness or even black colour. Such symptoms are not very specific because other factors than nematodes may cause similar effect. Even different nematode species may be responsible for similar types of discolourations. For example, *Aphelenchoides ritzemabosi* is responsible for interveinal discolouration in strawberry and chrysanthemum and a similar effect is produced by *Ditylenchus dipsaci* on aster leaves. Discolouration of stem is exemplified by *Rhadinaphelenchus cocophilus* on coconut and by *Bursaphelenchus xylophilus* on pine trees. On coconut, *R. cocophilus* causes production of a band of necrotic tissues in the lower part of the stem followed by appearance of reddish colour as a result of death of cells.

7. *Stem Galls.* Many species of *Anguina* from galls on stem. Examples are, *A. tumafasciens* on *Cynodon transvaalensis* (greenish or reddish galls) ; *A. australis* on *Erharta longiflora* *A. pustulicola* on grasses (pustule like galls) ; *A. cecidoplastes* on *Andropogon pertusus* ; *A. picridis* on *Aeroptilon picris* (orchid) ; *A. amsinckia* on *Amsinckia intermedia.*

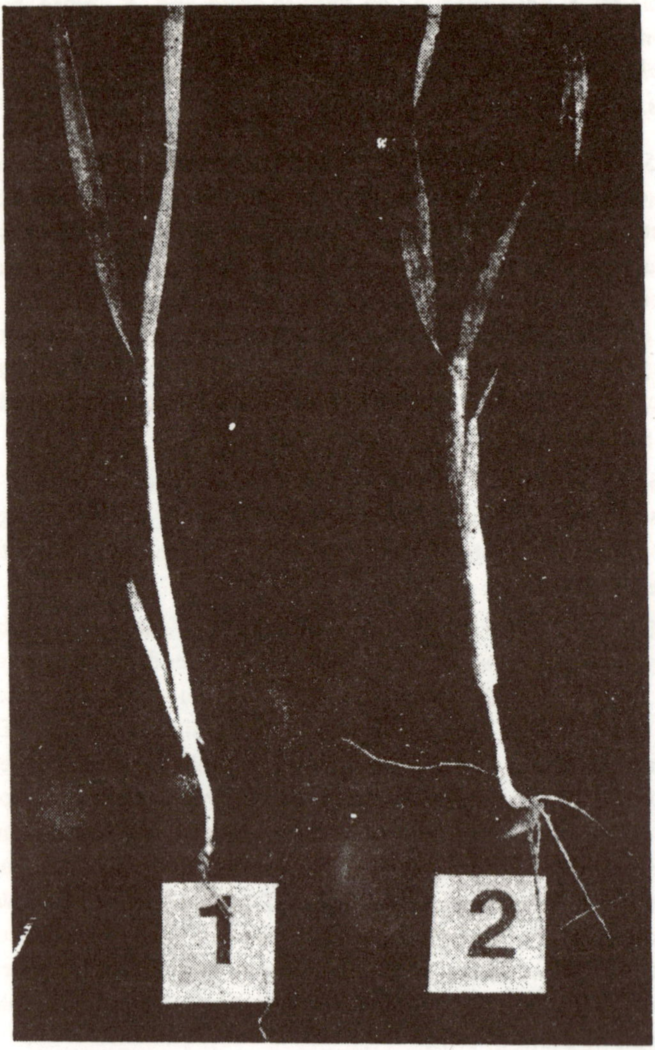

Fig. 4. Wheat seedling (right) showing basal swelling caused by *Anguina tritici*.

Left—Healthy seedling.

Abnormal growth in stems and leaves, as a result of hypertrophy or hyperplasia of the cells, may produce swellings, which are often diagnostic. *Anguina tritici*, soon after infection produces swelling near the basal part of the seedlings (about three week old seedlings) (Fig. 4) which gives the first indication of infection. Swellings in the infected tissues are commonly encountered with *Ditylenchus dipsaci* and *Aphelenchoides ritzemabosi* infestations on narcissus and strawberries respectively.

Below Ground Symptoms. As in the above ground symptoms, there is a wide range of symptom expression on the underground parts of the plants which sometimes are diagnostic but many times overlap or are nonspecific type.

1. *Root Galls.* These are characteristically produced by the root-knot nematodes (*Meloidogyne* species) (Fig. 5), false root-knot nematode (*Nacobbus* species) and small galls on roots of oats, barley, rye and wheat by *Ditylenchus rudicicola.* Small root galls can be produced also by *Hemicycliophora arenaria* on citrus and *Xiphinema diversicaudatum* on roses. The size and number of root galls may vary with host crops and the nematode species. *M. incognita* parasitisation on vegetable crops, especially cucurbits, results in formation of big root galls (Fig. 6) while the same species on cereal crops like wheat produces very small galls.

2. *Root Rots.* (Fig. 7) Many nematodes parasitizing fleshy roots serve to incite and aggravate activity of many other microorganisms. Once the nematodes have initiated the damage on root system, other scondary micro-organisms enter through the injuries causing extensive tissue destruction, which ultimately results in root rotting. Infestation of potato tubers by *Ditylenchus destructor* is first characterised by small discoloured superficial spots which may spread throughout the tuber surface. This is followed by drying and shrinking and cracking of skin. Micro-organisms gain entry through these cracks, continue their activity during storage and cause rot of the tubers. The yam nematode, *Scutellonema bradys* offers another excellent example of this type of nematode damage.

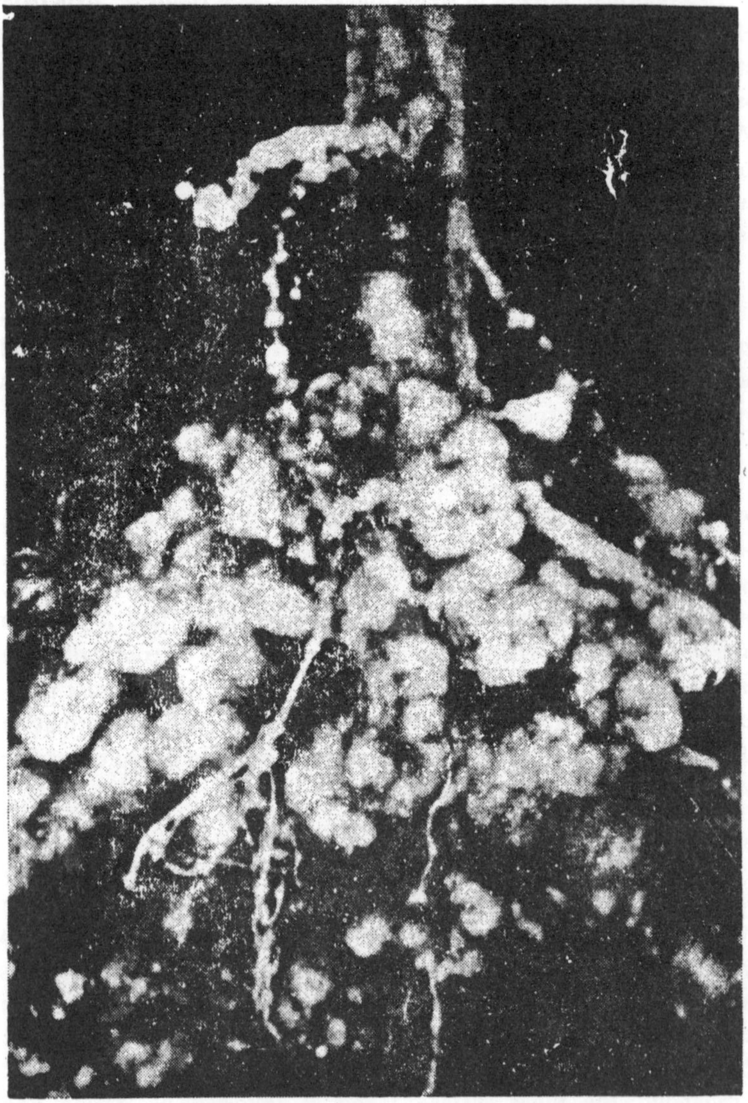

Fig. 5. Root galls caused by *Meloidogyne* species.

Fig. 6. Root galls on cucurbits caused by *Meloidogyne incognita*.

Fig. 7. Potato root rot caused by *Ditylenchus destructor*.

3. *Lesions*. These are, more or less, distinctly circumscribed necrotic areas. Typical lesions are produced by the lesion nematodes, species of *Pratylenchus* burrowing nematode, *Radopholus similis* (Fig. 8) and rice root nematode, *Hirschmanniella* species. Sometimes external feeders like species of *Criconemoides* may also produce small lesions.

4. *Necrosis*. Ectoparasitic nematodes, in large numbers, may cause superficial discolouration due to killing of cells over large areas. *Xiphinema* species, *Tylenchulus semipenetrans*, *Aphelenchoides parietinus* are often recorded associated with this type of injury.

5. *Excessive Root Branching*. Invasion of young roots by nematodes often stimultates development of root proliferation near the point of invasion. Some of the economically important nematodes like root-knot, cyst formers, lesion nematodes and *Nacobbus* species cause this type of symptom which has been variously referred to as 'witche's broom', 'hairy root' or 'bearding'.

Fig. 8. Coconut roots showing lesions by *Radopholus similis.*

Fig. 9. Onion seedlings showing bloat symptom caused by *Ditylenchus dipsaci*.
Healthy seedlings at extreme left.

6. *Devitalised Root Tips.* Many nematodes penetrate roots just behind the root tip resulting in stoppage of further growth and a situation arises when the root system appears to be composed of short clusters of coarse or stubby roots as with species of *Belonolaimus* and *Trichodorus* respectively. Abrupt halt in growth of lateral roots just as they are emerging from the main root results in the appearance of the main root as devoid of small rootlets.

Some nematodes like species of *Xiphinema* (in some cases) or even root-knot nematodes, may feed very near the root tip causing not only retardation of root growth but also giving a curl to the tip area, referred to as 'curly tip'.

It is very seldom that a single type of symptom can be correlated directly with a nematode species. Most often, the infested plants exhibit a combination of symptoms which sometimes overlap

Fig. 10. Yellow ear rot symptoms on wheta ears caused by *Anguina tritici* and *Corynebacterium michiganense* pv *tritici* complex.

in many nematode species. Distortion or twisting of leaves are produced by *Anguina tritici* as well as *Ditylenchus dipsaci* but since the hosts are different in these two cases, therefore, these symptoms become diagnostic. The same nematode species may produce different symptoms on different hosts. An excellent example is that of *D. dispsaci* which produces a symptom 'bloat' in onions (Fig. 9) and 'spikkels' on narcissus.

Another facet of symptomatology involving nematodes is in disease complexes with other micro-organisms like fungi, bacteria, viruses, etc. In addition to providing suitable avenues for entry of other organisms, otherwise weakly pathogenic, the nematode infestation may alter the internal plant substrate favouring growth of the other organisms. Besides these opportunistic complexes, there are often situations in which nematode plays an essential part in the disease complex, either as an aggravator or resistance breaker. The yellow ear rot disease of wheat is reproducible only if the nematode larvae are involved in the inoculations. The bacterium (*Corynebacterium michiganense* pv. *tritici*) spores are carried externally by the nematode larvae to the growing point of wheat seedlings and only then the bacterium is able to multiply, under proper environmental conditions, and produce the yellow ear rot symptoms (Fig. 10). The cauliflower disease of strawberry is another example where the nematode, *Aphelenchoides fragariae* is involved with the bacterium, *Corynebacterium fascians*.

In the soil environment, plant parasitic nematodes are one of the several biotic factors which, directly or indirectly, influence plant health. Mere presence of nematodes in the rhizosphere of unhealthy looking plants may not be conclusive enough to implicate nematodes unless pathogenicity tests are carried out to define the role of the nematodes, if any. However, in some casses, as stated already, the symptoms are diagnostic but still the overall role of the nematode in the plant health would need to be established to correlate damage to the nematode populations.

7

Genus Anguina Scopoli, 1777

The type species, *Anguina tritici* (Steinbuch, 1799) Filipjev, 1936, is the first authentic record of a plant parasitic nematode. These are typically gall forming nematodes of aerial parts of plants. *A. tritici* is the most important species attacking wheat. Barley is another host for the nematode, though field occurrence is rare[1]. Amongst the other species, *A. agrostis* (Steinbuch, 1799) Filipjev, 1936, *A. graminis* (Hardy, 1850) Filipjev, 1936 and *A. balsamophila* (Thorne, 1926) Filipjev, 1936, are of some importance.

Anguina tritici (Steinbuch. 1799) Filipjev, 1936, the wheat gall nematode.

The ear-cockle disease of wheat was discovered by a Dutch clergyman, Needham, in 1743[2] who first mistook it for smutted grain of wheat, but on examination, observed larvae within the galls. Subsequently, the causal relationship between the disease and the worms was demonstrated[3]. Later, it was named as *Vibrio tritici*[4]. The other names given during the course of time were ; *Rhabditis tritici, Anguilulla tritici, A. graminearum, Anguilullina tritici, Tylenchus tritici, Anguillula scandens* and *Tylenchus scandens* The genus *Anguina* erected in 1777[5] was revived in 1936[6] and the name *A. tritici* was established.

Morphology

The females (3-5 mm) are obese, generally assuming the shape of a closed spiral when relaxed by gentle heat. Sometimes, they are ventrally arcuate or with 'S' shape, broad in the middle and abruptly tapering towards anterior and posterior ends. Ovary is greatly developed with two or more flexures at its anterior end. Anterior end of the uterus is swollen into receptaculum seminis and the posterior end leads into a post-vulval uterine sac. Tail tapers to an acuminate end. In some cases, bifurcation of tail tip or variations in the tail tips can be seen also (Fig. 1).

The male (1.9—2.5 mm) has a straight and vermiform body with testis having one or two flexures. The bursa does not envelop the tail completely. The tail shape may vary greatly and may have also bifurcated tip.

Distribution

The nematode galls have been disseminated through infested/ contaminated seed to practically all wheat producing regions of the world and at one time or the other has been reported from most of the European countries, Asiatic region of the USSR, Israel, Syria, China, Pakistan, Australia, Newzealand, Egypt, Brazil, India, USA. The nematode galls have been referred to under various names in different countries. Purples, ear-cockles and pepper corns are the common names for England ; radenkrankheit for Germany ; blenielle for France ; eelworm disease, cockle wheat, nematode galls for USA ; and *sehr* . for India. The bacterial disease of wheat, commonly associated with it, is popularly known as 'yellow slimy disease' or in vernacular as '*tundu*' or '*tannan*'.

Biology

Nematode galls which fall in soil alongwith the contaminated seed at the time of sowing or may be present already from the previous season, absorb the soil moisture and become soft, facilitating the release of the second stage larvae. Germination of wheat seed is always preceded by release of the larvae in the soil[7]. Normally, wheat seed starts sprouting within 3 days after planting and the earliest release of larvae from galls is on the 4th day[8]. The larvae enter the seed, both from the brush as well as the embryonic ends.

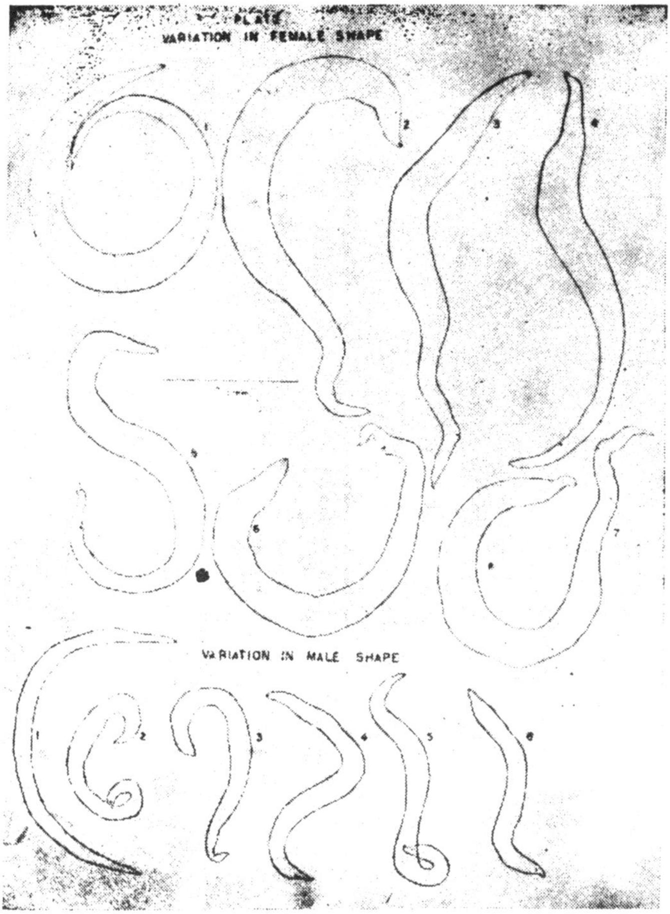

Fig. 1. Variations in female and male shapes of *Anguina tritici.*

However, the larvae entering the brush end are not able to reach the embryonic area from this source. The larvae entering from the embryonic end reach the embryo within 6 days and thereafter infest the growing point and are carried mechanically upwards with the growth of the seedling. While on the growing point, the larvae feed ectoparasitically[9] and enter endoparasitic mode of life, after about 90 days, invading the floral primordia developing on the

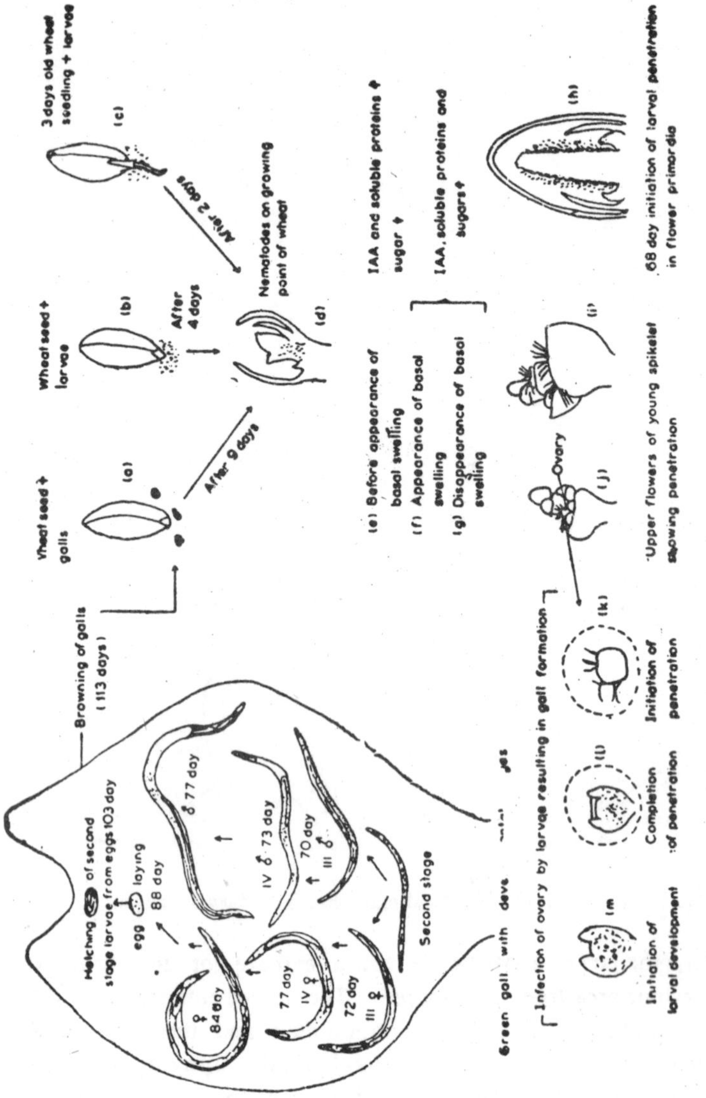

FIG 2 Diagrammatic representation of life stages of Anguina tritici

apical meristem. Within 3-5 days, the second-stage larvae, in the floral tissues, become adult males and females through 3rd and 4th moults. The adult females lay eggs within 6-10 days and eggs in various developmental stages as well as the next generation second-stage larvae, along with males and females, are noticeable during the next 15-20 days at the green gall stage. Subsequently, the green galls become brown and later change to black hard galls. These galls may fall in the soil at harvest and may serve as the source of new infection in the following season, if wheat is taken in the same field. However, majority of the galls remain as contaminant of the harvested seed and serve as the source of new infestation with the use of the contaminated seed for sowing. The total life cycle period, from second-stage larvae to second generation second-stage larvae is between 105-113 days[10]. A diagrammatic presentation of life cycle is given in Fig. 2.

Occasionally, the second-stage larvae present on the growing point of the seedlings may enter the leaf tissues in the early stages or enter the glumes or awn tissues at the flowering time. Small galls may be formed in these tissues also but the number of larvae present in such galls is few. A leaf gall may contain 25 nematodes which may be in various developmental stages and may include egg-laying females also[11].

The number of larvae present in the black hard galls ranges between 3000 to 30,000 depending upon the size of the galls. The total number of adults as well as the sex ratio is also affected by the size of the galls. In large galls (4-6 mm), the maximum number of adults recorded was 81 comprising of 55 females and 26 males. In small galls (upto 2 mm) this figure was 11 (7 females and 4 males) and in medium sized galls (2-4 mm), the number was 24 (16 females and 8 males). Irrespective of the size of galls, the female : male ratio was 2:1[12]. However, the number of larvae "in a very large gall" has been reported upto 90,000[13].

The viability of the larvae present in the nematode galls is reported to be quite long, though actual experiments have not been conducted on this aspect. Live nematodes from galls stored under

Fig. 3. Basal swelling, near soil base in wheat seedling infected with *Anguina tritici*.

dry conditions for 28 years have been recorded[14]. According to some other reports, juveniles in the galls can survive an exposure at —40°C for five minutes and ice embedding for several days [11,15].

Symptoms

Enlargement of the basal stem part, near the soil base, is the first visible symptom that can be seen in 20-25 days old seedlings[16] (Fig. 3). The leaves, emerging from such seedlings, are twisted and crinkled (Fig. 4). Frequently, some leaves are observed to be folded with their tips held near the growing point. Such leaves, however, recover almost to normal, 30 to 45 days after they have emerged from the growing point. In comparison to the healthy seedlings, the affected ones are dwarf with spreading habit. These symptoms are more apparent on young seedlings than those approaching maturity. At times, the seedling symptoms may not be very apparent though the plants may ultimately show some galls in the earheads.

The infected plants have more number of tillers with a comparatively faster growth rate than the healthy ones. This may give an impression that the infested plants are giving a better plant stand than the healthy ones (Fig. 5). There is an early information of ears in nematode infested plants, may be about 30-40 days in advance of their emergence in healthy ones. The ears from infected plants are either slightly or distinctly different from the healthy ears, depending upon the intensity of gall formation. Generally, the affected ears are short and broad, with very short or no awns on the glumes (Figs. 6, 7). In such ears, either all or some of the grains are replaced with the nematode gall. Each spikelet may contain 1 to 5 galls. Frequently, galls may be formed at the base of awns and on the glumes as well (Fig. 8).

In the yellow slime disease (nematode and bacterial complex) (*tundu*), the characteristic symptom is production of a bright yellow slime or gum on the leaf surface of young plants and also ears and leaves in contact with them while they are still in the boot leaf. Yellow slime, which can be seen trickling down the tips, under humid conditions, becomes hard, brittle and brown in colour on drying. The infected culms either die in the young stage of growth or may grow until the heading stage. In the latter case, either the emerging spike is narrow or short with grains partially or completely replaced by the bacterial mass or it may fail to emerge within the boot leaf (Fig. 7). The stalk is always distorted when such infected ears show the bacterial infection.

One of the common observations for the nematode infested wheat plants is that they are dwarfed, have more tillers and in the initial stages, the growth rate is comparatively more than the healthy plants[16]. An evaluation of the gall extract has demonstrated

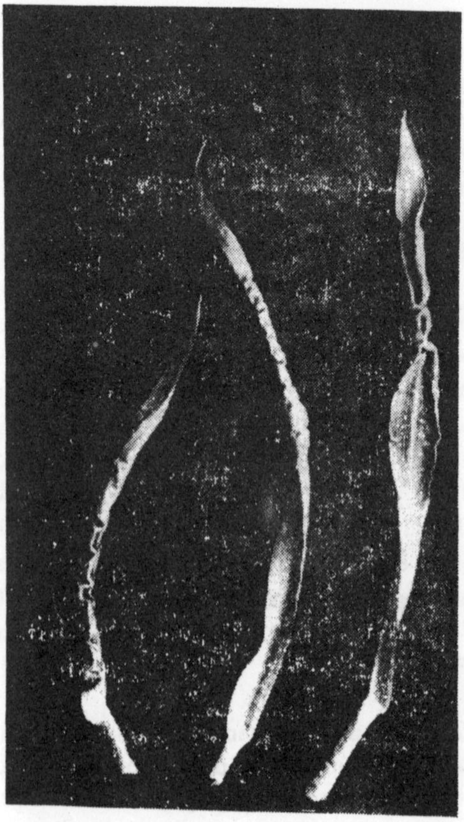

Fig. 4. Twisting and crinkling of leaves emerging from *Anguina tritici* infected wheat seedlings.

Fig. 5. *Anguina tritici* infected plants (right) exhibiting better plant stand than healthy (left) ones.

Fig. 6. *Anguina tritici* infected wheat ear-heads. Healthy earhead at extreme left.

that there is stimulation of the seedlings between 3 and 6 weeks of planting but this stimulation diminishes with the ageing of the seedlings. The gall extracts, on analysis, have shown presence of amino acids (aspartic acid, glutamic acid, glycine serine, tyrosine, alanine, tryptophan, lysine, nor-leucine, isoleucine, phenylalanine and proline) equivalent to 8 micromoles of leucine, in one g extract. Sugars equivalent to 42 mg glucose are also recorded[17].

Fig. 7. Variations in *Anguina tritici* symptom expression in earheads in some wheat varieties. Three earheads at extreme right exhibiting yellow ear rot symptoms caused by *A. tritici* and *Corynebacterium michiganense* pv. *tritici* complex.

Toxicity of Galls

There is no evidence, so far, which suggests the nematode galls have any mamallian toxicity. White albino rats fed with a mixture of basal diet and crushed nematode galls in the ratio of 9:1 did not show any adverse effect on the rats (Table 1). In fact, the rats consumed more quantity of the diet which was mixed with nematode galls in comparison to the basal diet alone. The rats also gained in weight[17]. However, another species of *Anguina*, *A. agrostis* has been reported to be toxic to animals[18].

TABLE 1

Diet Consumed and Weight Gained by Rats in 13 Days

Diet	Rats	Initial Weight (g)	Diet Consumed (g)	Weight of Rats After 13 days	Loss of Grain in Weight (g)
Basal	Female	123.1	63.0	67.5	+4.5
	Male	115.1	70.0	67.5	—2.5
Mixed with Galls	Female	160.3	73.0	80.0	+7.0
	Male	149.8	65.0	70.5	+5.5

Effect of Nematode Galls on Wheat Flour

Since the nematode galls remain as a contaminant with wheat grain, it is quite possible that the galls are also crushed along with the wheat grain at the time of crushing. A comparison of the crude protein, pelshenke value, etc., of the wheat flour with nematode gall flour and also in different ratio combinations of wheat and nematode gall flours has shown that the wheat flour has 11.5 per cent protein as against 21.7 in nematode gall flour. When the wheat flour is mixed with nematode gall flour in the ratios of 4:1, 3:2, 2:2 and

1:4, the mixtures possess the protein percentages as 13.0, 15.1, 17.4 and 19.2 respectively. Gluten is absent in the gall flour. The Pelshenke value of the wheat flour was 140 but when mixed with the gall flour in the above ratios, it gradually fell to zero in the gall flour. In the 'chapati' making test, the wheat flour could tolerate a contamination with gall flour upto 20 per cent, as far as handling qualities were concerned. With regard to the baking qualities (puffing, keeping, appearance, colour, texture, smell and taste), it could tolerate a contamination only upto 5 per cent (by weight)[19].

Physiologic Specialization

Till recently, there was no direct evidence of presence of any physiologic specialization either in the nematode population or in the bacterium associated with it. However, considerable variations in the shape and size of nematode body were recorded[19] and it was suggested that these variations could be because of geographical regions. In a recent study, five different geographical populations

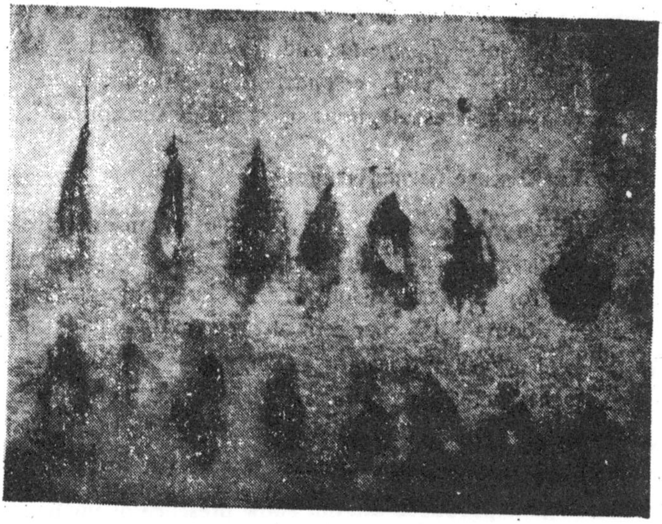

Fig. 8. Nematode gall formation at the base of awns.

were evaluated under Indian conditions, on six wheat varieties and it was observed that there were two distinct populations, both with regard to nematode reproductivity and reactions on the wheat cultivars[10].

Loss

In the early twenties, losses as high as 70 per cent have been reported[15] from wheat growing regions. Leukel[20] recorded an outbreak of this disease in the Virginia State of the United States of America, as a result of continuous wheat culture without rotation in response to a campaign 'raise more wheat and win the war'. Many farms were so badly infested that they could not be harvested at all. The disease is still a problem in the wheat regions of India. According to an estimate, yield losses in one State (Delhi) alone ranged from 1-3 per cent, with as much as 50 per cent in individual cases[21].

Hosts

Under field conditions, wheat (*Triticum aestivum*) is the only host. Recently barley (*Hordeum vulgare*) has been recorded also as a host, though a poor one[1]. *T. spelta* and *T. dicoccum* and rye are recorded as susceptible[11,15,22]. *T. compactum* is also recorded as susceptible[17]. Amongst grasses, *Aegilops* sp. is highly susceptible[20].

The Yellow Ear Rot Disease (*tundu*/tannan)

Corynebacterium michiganense pv. *tritici* (Hutchinson (Dye & Kemp (*C.m.* pv. *tritici*), present in the nematode galls, is the primary source of inoculum for the expression of the bacterial symptoms. It is recorded that about 40-55 per cent of the nematode galls carry the bacterium[23]. However, it does not mean that some galls definitely do not carry the bacterium. The non-isolation of the bacterium from some of the galls could be either due to absence or limitations of the techniques employed in the studies. In an experiment, where surface sterilized larvae were used for inoculations of wheat seedlings, the nematode galls produced did not show presence of the bacterium, irrespective of autoclaved or non-autoclaved soil used in the experiments. An obligate etiological relationship between the nematode and the bacterium has been suggested[23-25]. In a later study, a positive association of *C.michiganen:e* pv *tritici* with nematode larvae has been

Fig. 9. *Corynebacterium michiganense* pv *tritici* spores attached to larval body of *Anguina tritici*.

demonstrated[10]. The bacterial spores remain attached with the external body surface of the second-stage larvae (Fig. 9), which on penetration to the growing point of wheat seed, carry the bacterial spores along with them. The question of association of bacteria with nematode galls is not confined to *C.m.* pv. *tritici* only. Several other bacterial types are also present in these galls[10]. However, there is no evidence that these other bacterial types play any role in the disease complex.

The type of the inoculum as well as the soil, determines the symptom expression of *tundu* or ear-cockle. It has been established that if unsterilized larvae are used for inoculations in unsterilized soil, maximum expression is for ear-cockle (74.6 per cent) with a mean number of galls as 135. However, with unsterilised larvae in sterilised soil, there is a comparative increase in *tundu* expression (20.5 per cent as against 2.0 per cent), and it was suggested that this could be because of the possibility that the bacterial colonies on the surface body of larvae are protected from interference from other soil microflora. With the use of unsterilized galls as the inoculum source, the percentages for ear-cockle and *tundu* symptom expression were 67 and 18 respectively[23].

In an attempt to understand the physiological basis of association of the bacterium and the nematode, nutritional requirements of the bacterium were evaluated. It was found that the gall extract contained all the essential amino acids (hystidine, glycine, aspartic acid and glutamic acid, which are required for the growth of the bacterium) and it was suggested that for the growth of the bacterium, the nematode galls contained adequate nutritional requirements[17].

Control

Nematode galls are the primary source of perpetuation of the disease and therefore, cleaning the seed lot, from nematode galls, is the most effective method of control. For this, mechanical cleaning or soaking the contaminated seed lot in 20 per cent brine solution, which allows the nematode galls to float up and thus can be skimmed off, can be followed. Care has to be taken to wash the seed lot with plain water 2-3 times after brine treatment, to remove the adhering salt particles, as otherwise seed germination is impaired. The brine solution is prepared by dissolving 40 lbs of sodium chloride (common salt) + a small quantity of potassium chloride in about 60 litres of water. The contaminated seed in the brine solution are stirred vigorously and frequently and the floating galls are skimmed off.

Mechanical separators have been devised also and have been used with success (99 per cent) in China[26].

At harvest, some of the galls may fall in the soil and may carry the infestation in the next season. Therefore, rotation of field may help also. Under tropical humid conditions, the galls may rot away in the soil. Therefore, seed contamination remains the main source of infection. Some of the bacterial flora like *Bacillus subtilis*, *B. pumilis B. cereus*, *etc.*, obtained from some nematode galls, are reported to be larvicidal. Wheat seed treated with the suspension of these bacterial organisms have shown very little infestation by the nematode[10].

Association of the Nematode with Other Pathogens

Pycnospores as well as the secondary spores of a fungus (*Dilophospora alopecuri*) have been found attached to the larvae[27]. The larvae carried the fungal bodies to the growing point of the wheat seedlings. Initially, the fungal disease could not be reproduced in the absence of nematode larvae but later it was established that the fungus can cause the disease in wheat independently. Therefore, the nematode larvae may probably be functioning as a vector without any obligate relationship of the type found in the yellow ear rot disease. The disease appears to be of minor importance, confined to the temperate regions of the European countries.

Other Species of *Anguina*

The other species of *Anguina* are mainly restricted to grasses with only *A. agrostis* being of some importance because of the mammalian toxicity of the galls. Brief remarks on some important species are as below :

A. agrostis **(Steinbuch, 1799) Filipjev, 1936**

Commonly known as Bentgrass nematode ; parasite of grasses of the genus *Agrostis* mainly. Also infest species of *Festuca, Phleum* and *Poa*. Elongate galls are formed in the earheads. *Festuca* galls are known to be fatal to cattle, sheep, horses, etc., in U.S.A.

A. australis Steiner, 1940

Leaf galls of a grass, *Erharta longifolia* in Australia.

A. balsamophila **(Throne, 1926) Filipjev, 1936**

It causes leaf galls of balsam root, *Balsamorrhiza sagittata* in U.S.A.

A. graminis **(Hardy, 1850) Filipjev, 1936**

Also known as Fescue leaf gall nematode and parasitizes on *Festuca ovina* and *F. rubra*. It is recorded from Europe.

A. klebahni **Goffart, 1942**

It infests flowers of primrose, *Primula Florindae* in Germany.

A. microlaenae **(Fawcett, 1938) Steiner, 1940**

It causes galls on shoots, leaves and infloresence of *Microlaena stipoides* in Australia.

A. millefolii **(Low, 1874) Filipjev, 1936**

It produces leaf galls on *Achille millefolium* and *A. tanacetifolia*. It is recorded from European countries including England.

A. picridis **Kirjanova, 1944**

It has been recorded from U.S.S.R. on stem of an Orchid, *Acroptilon picris*.

***A. spormophaga* Steiner, 1937**

It is recorded from the infloresence of an important grass, *Saccharum spontaneum*, in U.S.A.

***A. tumafascians* Cobb, 1932**

On *Cynodon transvaalensis*, it produces ovoid galls on stem, leaves and sometimes in flower heads. It is known to be present in South Africa.

REFERENCES

1. Bhatti, D.S., Dahiya, R.S. and Dhawan, S.C., New record of *tundu* and ear-cockle in barley. *Nematologica*, 24, 331-332, 1978.

2. Needham, T., Concerning certain chalky tuburous concretions, called, malm ; with some microscopical observations on the fauna of the red lilly and of worms discovered in smutty corn. *Phil. Trans.*, 42, 634, 1743.

3. Roffredi, D.M., Memoire sur lorigine, des petitis vers. on anguilles du ble rachitique. *Observ. mem. phys. hist. nat.*, 5, 1-19, 1975.

4. Steinbuch, J.G., Das Grasalchen, *Vibrio agrostis. Naturforscher*, 28, 223-259, 1799.

5. Scopoli, J.A., Introductio and historiam naturalem sistens genera lapidum, plantarum, et animalium, Prague, 1777.

6. Filipjev, I.N., On the classification of Tylenchinae. Proc. helminth. Soc. Wash., 18, 110-112, 1936.

7. Carne, W.M.,. Ear-cockle (*Tylenchus tritici*) and a bacterial discease (*Pseudomonas tritici*). of wheat. *J. Dept. Agric*, W. Aust., 3, 508-512, 1926.

8. Pathak, K.N. and Swarup, G., Emergence of juveniles from *Anguina tritici* galls, numbers reaching growing point of wheat seedlings and relationship between inoculum levels and number and size of galls produced. *Indian J. Nematol.*, 13, 155-160, 1983.

9. Midha, S.K., Chatrath, M.S. and Swarup, G., On the feeding of *Anguina tritici* on growing point of wheat seedlings. *Indian J. Nematol.*, 1, 93-94, 1971.

10. Gokte, N., Studies on biology and host-parasite relationship of the ear-cockle nematode, *Anguina tritici* on wheat. Ph.D. Thesis, IARI, New Delhi, 1987.

11. Byars, L.P., The nematode disease of wheat caused by *Tylenchus tritici*. U.S. Dept. Agric. Bull., 842, 1920.

12. Midha, S.K. and Swarup, G., Studies on the wheat seed gall caused by *Anguina tritici*. *Indian J. Nematol.*, 4, 53-63, 1974.

13. Christie, J.R., Plant Nematodes, their bionomics and control. Univ. of Florida, Gainesville, Florida, 1959.

14. Fielding, M.J., Observations on the length of dormancy in certain plant infecting nematodes. *Proc. helminth. Soc. Wash.*, 18, 110-112, 1951.

15. Leukel, R.W., Investigations on the nematode disease of cereals caused by *Tylenchus tritici*. *J. Agr. Res.*, 27, 928-956, 1924.

16. Gupta, P. and Swarup, G., On the earcockle and yellow ear rot diseases of wheat, I. Symptoms and histopathology. *Indian Phytopath.*, 21, 318-322, 1968.

17. Midha, S.K., Studies on the earcockle nematode and the development of ear-cockle and *tundu* diseases of wheat. Ph.D. Thesis, IARI, New Delhi, 92 pp., 1969.

18. Shaw, H.M. and Muth, O.H., Some types of forage poisoning in Oregon cattle and sheep. *J. Amer. Vet. Med. Assoc.*, 114, 315-317, 1949.

19. Gupta, Pramila, Studies on the ear-cockle and *tundu* diseases of wheat. Ph.D. Thesis, IARI, New Delhi, 106 pp., 1966.

20. Leukel, R.W., Nematode disease of wheat and rye. *Farmer's Bull.* USDA No. 1607, 16 pp., 1957.

21. Vasudeva, R.S. and Hingorani, M.K., Tanan or *tundu* disease of wheat. *Indian Fmg.*, 2, 14-19, 1952.

22. Rademacher, B., Aufreten and Bekampfung des Weizenalchen (*Anguina tritici*) beim Dinkel (*Triticum spelta*) in Zusammenheng mit der Federbuschsporenkrankheit (*Dilophospora graminis*)

und eine Bakteriose. *Zeitschrift fur Pflanzenschuntz*, 57, 334-343, 1950.

23. Pathak, K.N., Studies on the relationship between ear-cockle nematode, *Anguina tritici* and *Corynebacterium tritici* in the *tundu* disease complex of wheat. Ph.D. Thesis, IARI, New Delhi, 66 pp., 1981.

24. Cheo, C.C., A note on the relation of nematode (*Tylenchus tritici*) to the development of bacterial disease of wheat caused by *Bacterium tritici* Hutchinson. *Ann. appl. Biol.*, 33, 446-448, 1946.

25. Sabet, K.A., On the source and mode of infection with the yellow slime disease of wheat. *Bull. Fac. Agric.*, Cairo Univ., 42, 1-15, 1954.

26. Chu, V.M., The Prevalence of wheat nematode in China and its control. *Phytopathology*, 35, 288-295, 1945.

27. Atanasoff, D., The Dilophospora disease of cereals. *Phytopathology*, 15, 1-40, 1925.

8

Genus Ditylenchus Filipjev, 1936

Characteristic thickening of leaf sheath and stem of rye, accompanied by wavy leaf growth and general stunting of plants led to the discovery of a diseased condition, described as 'stock'[1]. The causal agent was later found to be a plant parasitic nematode, described under the generic name, *Anguillula dipsaci*, infecting teasel, *Dipsacus fullonum*[2]. The subsequent years witnessed the description of the same nematode under different names, *Anguillulina dipsaci* var. *amsinckiae*, *Anguillula devastatrix*[1–4], etc. Apparently, the confusion arose, by and large, due to the failure of earlier workers to recognize the existence of physiological races within conspecific populations of *dipsaci* and the ability of the nematode to attack different crops with varied symptomatology and pathogenicity. Finally, the confusion was partially resolved by Filipjev[5] when he proposed the genus *Ditylenchus* with *D. dipsaci* as the type species.

Siddiqi[6] recognizes 52 valid species under the genus *Ditylenchus* in addition to *D. dipsaci*. Of these, *D. dipsaci*, *D. angustus*, *D. destructor* and *D. myceliophagus* are of economic importance.

Ditylenchus dipsaci (Kuhn, 1857) Filipjev, 1936, The Stem and Bulb Nematode

Morphology

The female body is almost straight when relaxed. It is commonly known as the stem and bulb nematode. It is a migratory endoparasite feeding on parenchymatous tissues of stems and bulbs of plants. Lip region is low, unstriated, slightly flattened with moderate sclerotization in the head region. The spear is about 10-12 μm long with distinct knobs. The oesophageal bulb butts onto or slightly overlaps the intestine with a small valve at their junction. Tail is conoid, with a sharply pointed terminus. A post-vulval sac is present which extends about half-way to the anus.

The male nematodes are similar to females with ventrally curved spicules.

Distribution

It is one of the few plant parasitic nematodes which has received world attention due to its potentiality for economic damage to many of important plant species, especially in temperate regions. Many crop growing areas, of both the hemispheres, have recorded its presence. Obviously, its spread has occurred through seed, plant material and cut flowers. Its distribution includes Algeria, Greece, U.K., U.S.A., Germany, Scandinavia, Sicily, Spain, New Zealand, Argentina, Bolivia, Peru, Japan and North America[1,7,8].

Host Range

It is known to attack over 450 plant species, under 40 odd families. Some of the plant species that are considered as favoured hosts are teasel, daffodil, narcissus, tulip, hyacinth, onion, garlic, carrots, potato, sugarbeet, turnip, rye, corn, lucerne, alfalfa, phlox and strawberry.

Biology

Life Cycle

One of the unique features in the life cycle is the pre-adult or fourth stage larva, sometimes known also as 'eelworm wool' or 'nematode wool' stage which is extremely resistant to desiccation. It may retain viability under dry conditions for many years. Although, other larval stages are capable also of causing infection to crops, under favourable conditions, the pre-adult (fourth stage) stage is the most important stage in the life history of the nematode.

The fourth-stage larvae enter plant tissue through open stomata and then moult. The total duration of life cycle ranges from 19-23 days at 15°C with four moults and four larval stages, the first moult being within the egg[7]. Sexes and developmental stages can be differentiated on the basis of position and nature of genital primordium and body dimensions. Reproduction is amphimictic. The female start depositing eggs four days after final moult and continue for 25-30 days. On clover and oat, the life cycle is completed between 24-39 and 20 days respectively[8].

The fourth-stage larva can enter into an anabiotic phase under adverse conditions. The anabiotic fourth-stage larva loses more than 99 per cent of its body water, yet maintains considerably reduced reversible metabolism[9]. Recent studies on ultrastructural changes in anabiotic larvae of the nematode explain the mechanism which aid in survival of *D. dipsaci*[10]. The anabiotic larvae maintain their structural integrity despite decrease in thickness of cuticle because of packing of layers and condensation of sacroplasm of muscle cell. Following rehydration, there is a lag phase of 2-3 hours during which the animal undergoes repair before recovery. The lipid droplets within intestinal cell coalesce and the cuticle increases in thickness. The sacroplasm expands with increase in spacing of thick filaments. The mitochondria swell, become more spherical and the cristae become less distinct. In spite of impaired mitochondrial function and low ATP levels, the initial phase of rehydration is accompanied by a burst of metabolic activity. It can be said that the success of the anabiotic larvae to survive and resume normal activity is due to their ability to undergo repair which may involve restoration of membrane damage, re-establishment of the ionic gradients essential for normal muscle and nerve function, prior to recovery.

Among the plant parasitic nematodes, *D. dipsaci* perhaps shows greatest extent of physiological variations. There are more than twenty biological races, differing in their host range. The races have been named after the preferred hosts[11]. Atleast 11 distinct races are known to occur in Europe, identifiable on the basis of host preferences[12]. Races considered to be of economic importance are : garlic, onion, tulip, oat, alfalfa, red clover and rye. Several of these races are polyphagous, capable of attacking diverse plant species, whle others have limited host range[11]. For example, the alfalfa, red clover and white clover races seem to be very specific for the named hosts. In contrast, rye, oat and onion races are polyphagous. The different populations of the same race may exhibit differences in host range and varied pathogenicity. The suggestion made earlier that many races are distinguishable morphologically and could represent true species[12], has not been supported by data from crossing experiments[11,13,16]. It is possible to have successful crossings between a number of these biological races and their progeny could exhibit different host preferences.

The evolution of these races still remains in the speculative phase. Monoculture of specific host may favour certain gene combinations and suppress others, resulting in a change in the pathogenic potential of a population[17]. The race genesis in the stem and bulb nematode is further explained on the basis of the work done with a garlic isolate of *D. dipsaci*[17]. Under aseptic conditions, the isolate reproduces on a wide range of host tissues, many of which are the key hosts reported for the biological races of the nematode. It has been theorized, on the basis of these observations, that the nominal species, *D. dipsaci* gene pool can be considered as wild type and polyphagous, with race deviations as part of gene continuum.

Temperature, moisture conditions, soil texture and accompanying plant growth are important soil factors and determine the activity of the nematode under natural environment. It is likely that all these factors are inter-related and their resultant interactions determine the survival, persistence and activity of the nematode under field conditions. The desiccated and anabiotic pre-adult nematodes are capable to resist temperature extremes. It has thus become rather essential to moist the anabiotic larvae, residing inside propagation bulbs, in order to achieve effective control of the nematode, since they exhibit greater tolerance to heat than their host plants[18-22]. In contrast, the nematode has been demonstrated to survive low temperature of —80°C[23]. About 16-24 per cent of anabiotic larvae can survive after exposures to —196°C for 24 hours or —150°C for 18 months[24].

A temperature range of 20-25°C has been recorded as the optimum for raising *D. dipsaci* and symptom expression in alfalfa seedlings under laboratoy conditions[25,26]. The optimum temperature for reproduction of the nematode on many plant species is reported to be 19°C[27]. The degree of mobility, symptom expression, extent of host tissue damage and resistance of host are considered to be the function of temperature at the time of infectivity[28-32]. Fewer nematodes invade tolerant Lahontan alfalfa than susceptible cultivar Grimm at 15.5°C[28]. But the susceptibility of Lahontan increases with the elevated temperature of 25-30°C. In a similar study, it was recorded that Lahontan alfalfa was less resistant to *D. dipsaci* at 15.5 and 21°C than at 11°C, indicating break-down in resistance with the increase in temperature[29].

Soil texture is also one of the important factors that govern the persistence of the nematode under field conditions. Heavy soils which contain more than 30 per cent clay tend to stabilize the population to about 40 nematodes per 500 g soil, a number sufficiently high to damage entire field plant population[33]. In contrast, sandy or light clay soils are unfavourable for multiplication of population.

The amount of moisture present in the atmosphere also play an important role in determining the activity of the nematode. In fact, temperature is of major importance primarily in its effect on relative humidity[34]. The fact, that with an advent of high temperature, the pathogenic symptoms on alfalfa are suppressed, reflects on the effect of thermal influence on humidity[34,35].

Other factors like light and host nutrition may influence also reproduction of *D. dipsaci*. Exclusion of light[26] and application of calcium respectively may increase and decrease [36,37], the reproduction of the nematode.

The extent of life span of anabiotic larvae is important from dissemination point of view. Pre-adult larvae survive in dried teasel for 23 years[38]. Marked differences in the survival ability of the nematodes, recovered from different host sources, has been observed also. Anabiotic larvae, from onion can survive in dry soil for two years[39] and a year and half from carrots, without hosts. The extreme case is of survival for 7 years in moist soil without a host[40].

Fig. 1. Damage symptoms of *Ditylenchus dipsaci* on lucerne seedling.
Left—Healthy seedling ; Right—Damaged seedling.
(After Hague, 1978).

Freezing soils usually contain very few nematodes since there is a high mortality during severe winter. Nevertheless, *D. dipsaci* can overwinter in soil as adult or fourth-stage in absence of host[41].

Symptoms

Ditylenchus dipsaci is one of the few plant parasitic nematodes which cause severe damage to many crops of economic importance under temperate, cool and humid conditions. In France, it is considered as the second most important parasite of legumes with severe losses to clover and lucerne[42] (Fig. 1). Other crops like pea, beans, strawberry, etc., are also affected[42,43]. Under field conditions, the forage yield of nematode infected alfalfa fields have been found to be significantly lower than that of non-infected plants, at each harvest. Also the stored carbohydrates in the infected plants are significantly lower than that of non-infected plants.[44]

The symptom expressions, in infected hosts, vary with plant species involved and environmental factors. Under field conditions, the specific symptom expression of nematode infected plants of alfalfa is noticeable during re-growth, in early spring. The infected plants are stunted ; frequently, have a 'halo' of dead stem around them.[11] The developing shoots of infected plants are swollen with

Fig. 2. Damage symptoms of *Ditylenchus dipsaci* on onion seedlings.
Left—Healthy seedlings.
(After U.C., Riverside slide).

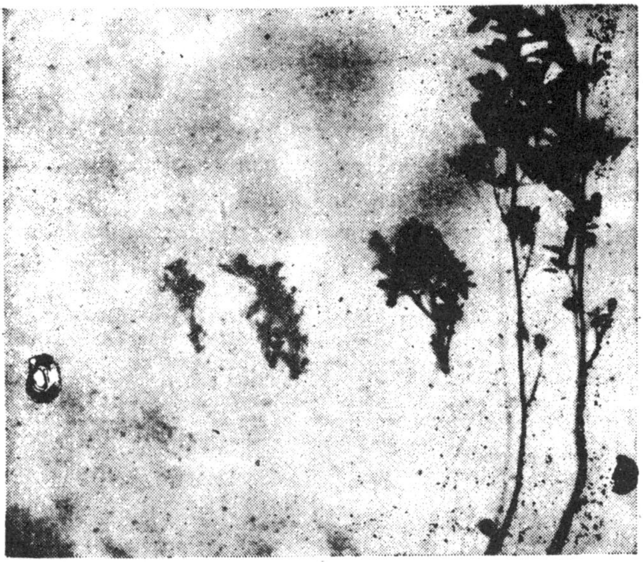

Fig. 3. Damage symptoms of *Ditylenchus dipsaci* on alfalfa seedlings. Healthy seedling on extreme right. After U C , Riverside slide).

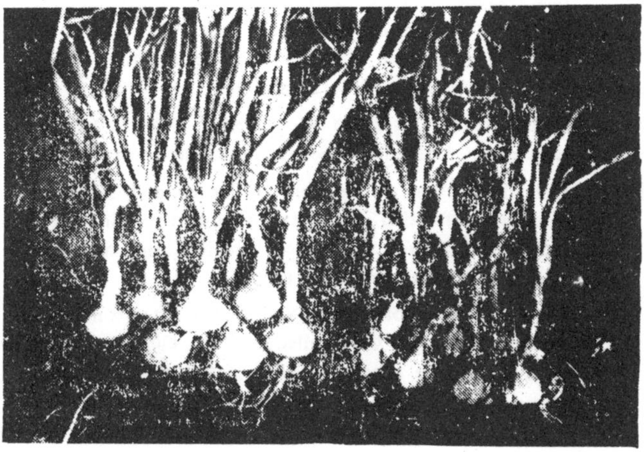

Fig. 4. *Ditylenchus dipsaci* damage to onion seedlings. Left—Healthy, Right—Diseased. (After U.C., Riverside slide)

shorter node. The leaves on the shoots are distorted and usually, but not always, remain folded (Fig. 3). The rapidly growing shoots may carry the nematode with them as a result of which swelling may be seen high up the stem. Nematode-infected plants, in established fields, may or may not recover with the advent of hot dry weather. The nature of the symptom expression in few important crops are given briefly in Table 1.

TABLE 1

Symptom expression in crops infected with *Ditylenchus dipsaci*

Crop	Symptoms
Teasel	Dwarf 'puff ball' type deformity due to infection on woody vascular tissue of burs.
Narcissus	Leaves of severely infected bulbs bear small elongate swelling known as 'spikkles'. In advanced stage, swelling may become yellow or brown in the centre ; brown ring may be present in infected bulb. Partially infected bulbs may exhibit stunted leaves and flowers or fail to produce flower.
Onion	Infected seedlings twisted, deformed and frequently die. Infected plants irregular in form with short and thickened leaves. Softening of bulbs occurs at the neck, which proceeds downwards causing bloat. In presence of moisture, soft rot sets in with offensive odour (Figs. 2 and 4).
Red clover	Deformed buds ; stunted shoots with swollen base and stem with galls. Loss of foliage leading to the death of plants.
Strawberry	Gall like swelling at the base of leaves and stems and upto petioles and fruiting branches. Frequently yellow, brown, or red spots present ; distorted, swollen and twisted foliage. Foliage and fruit production reduced.
Phlox	Terminal portion of growing young stem often transformed into elongate swelling. On swollen portions, closely arranged leaves form a bunchy rosette. Taller stems may be spindly with elongate nodes, bearing deformed and contorted leaves with irregular edges. Severely infected plants with large number of deformed and short stems giving witches'—broom effect.
Potato	Swelling and crinkling of stems and leaves.

The effect of moisture, under experimental conditions, on symptom expression, is rather remarkable. In alfalfa, the nematode enters the emerging seedings at the base of cotyledonary petioles. Subsequently, cavities are formed in the cortical parenchyma within 12 hours of infection.[46] Characteristic swellings are observed on the stem within next 24 hours. Depending on the degree of humidity that prevails in the atmosphere, the infected plants appear to be short and distorted. Continued low humidity even in the presence of sufficient moisture and moderate infection level may lead to some plant mortality within a few weeks.

Histopathologically, galled shoots become spongy and are easily broken. Complete destruction of cortcal parenchyma, withdrawal of cytoplasm, cell wall collapse, formation of cavity, enlargement of cell and loss of chloroplast are some of characteristic changes exhibited by infected plant parts[46-50]. In alfalfa tissues, infection results in swollen and broken endoplasmic reticulum, distended and broken chloroplast, loss of nuclear material and bulging or rupturing of nuclear envelope. Cytoplasm of infected cell show more endoplasm, ribosome vesicles and golgy apparatus indicating enhanced metabolic activities.[51]

Conspecific population of *D. dipsaci* exhibit differences in the quality and quantity of some hydrolytic enzymes but no apparent correlation can be detected between enzyme activity and differing pathogenesis[53]. Pectinase activity has been reported from various populations of *D. dipsaci*, raised on different hosts. The survival, multiplication and successful feeding in host depends on its ability to macerate host tissue through dissolution of middle lamellae enzymatically, without the death of cells. Because pectic compounds are considered to be important structural components of middle lamellae, it is presumed that pectolytic enzymes are instrumental as well as essential for dissolution of middle lamellae, which, in the process, can function as a factor for pathogenesis. A variety of pectolytic enzymes from *D. dipsaci* has been reported from different laboratories. Although earlier work did not provide conclusive evidences for the involvement of the nematode to induce pathogenesis through pectolytic enzymes[52],[53], convincing results[54] are now available which demonstrate the implication of pectolytic enzymes in the symptom expression of infected hosts. Presence of two pectolytic enzymes in aqueous extractions of *D. dipsaci* have been demonstrated[54]. The enzyme, identified as endopolygalacturonase (endo-PG) is responsible for the separation of the host cell and it is of nematode origin and not of host.

The pectolytic enzymes may differ qualitatively as well as quantitatively in different populations of *D. dipsaci*. These differences may be due to procedural techniques for assays or source of the host plant-nematode culture[54]. Perhaps, the host specificity is related to the differences in the kind of the enzyme activity, and could induce fundamentally different mechanisms of pathogenesis.

In association with other micro-organisms, the nematode is suspected to enhance, in certain cases, severity of plant diseases. The increase in pathogenic expression of rhubarb crown rot, a bacterial disease caused by *Bacterium rhaponticum*, was suspected as early as 1936[55] because of association of *D. dipsaci* with the disease. It has been also reported to transmit *Corynebacterium insidiosum* and pre-dispose wilt resistant alfalfa to infection[56]. *D. dipsaci* and *Pseudomonas fluorescens* are associated also in 'Caf au lait', a disease of garlic[57].

Antagonistic type of relation has been demonstrated between *D. dipsaci* and tobacco mosaic and tobacco rattle viruses. Such associations result in reduced reproduction of the nematode. Contrary to this, plants infected with arabis mosaic and black ring viruses influence the nematode reproduction favourably[58].

Control

The control of stem and bulb nematode is based mainly on basic principles which emphasize the use of clean planting stock and denematization of propogative materials, either through fumigation or hot water treatment. Methyl bromide fumigation of seed and bulb at atmospheric pressure of 70 g/m³ for 20 hours (CT=1400), 80 g/m³ for 16 hours (CT=1280) and in a vacuum at 50 g/m³ for 20 hours, 60 g/m³ for 16 hours, 100 g/m³ for 8 hours (CT=1000, 980 and 800 respectively) has proved successful[59]. Germination of planting material is reportedly not affected even two years after treatment. In the past, DBCP, applied at the time of alfalfa seeding, was found quite effective in reducing reproduction of the nematode. Good to excellent control of the nematode has been obtained with non-fumigant chemicals, like aldicarb on narcissus, onion, tulip ; carbofuran on narcissus, onion ; dazomet on onion ; demeton on narcissus ; oxamyl on narcissus and onion ; phenamiphos on narcissus and onion ; phorate on narcissus and onion ; scharadan on narcissus ; thionazin on narcissus, onion, tulip, oat, hydrangea, red clover and white clover ; and Shell OS 1836 on narcissus and tulip[60]. Potassium metabisulfate tablets (at 400 mg/1000 cm^3 distilled water) have provided effective control of the nematode in onion and garlic[61].

Infected seeds and bulbs of several plant species can be successfully treated for denematization by immersing them in hot water. In general, exposure of dormant bulbs to hot water bath at 44-45°C for 3 hours, yields satisfactory results. Pre-soaking plus wetting agent and addition of fungicides aid in better control of the nematode. Soaking at 100°F for 30 minutes in a formalin (1 per cent)—detergent solution (0.1 per cent), followed by a treatment for 20 minutes at 120°F in the same solution eradicates the nematode in garlic cloves.

Crop rotation is also an effective method and has been recommended for many crops like oat, rhubarb, strawberry clover, narcissus, alfalfa, onion and teasel. The success of crop rotation, as a primary means for the control of the nematode, depends on knowledge about existence of biological races, host range including crops and weed hosts, environmental factors like temperature, relative humidity and soil types. A rotation of 3-10 years, with non-host crops, has been recommended for teasel, narcissus, onion and alfalfa[60,62,63].

Sources of resistance against the nematode have been identified in some cases (Table 2). Alfalfa is one of the several crops which has been extensively studied for incorporation of stem nematode resistance in commercial cultivars. Employing original source of resistance from Turkey or Iran, nematode resistant lines, Turkestan, Nemastan, Lahontan, Keyseri and Talent have been developed

TABLE 2

Sources of resistance in crops against *Ditylenchus dipsaci*

Crops	Lines/Variety	Reference
Alfalfa	Lahontan	Hunt *et al.*[66]
	Falkiner	Moore and McLeod[67]
	Nematol, Alfa 11 Vertus	Bingefors[68]
Broad Red Clover	Peares	Brown[69]
	Dorset Marl	Fiddian and Aldrich[70]
Red Clover	Merkur, Resistenta, Reko, Disa, Ulva	Bingefors[68]
	Maris leda, Sabotoron	Howard and Cotten[72]
White Clover	Mira, Mega, Trifo Zero, Lodi, Adera, Milka, Milkanova	Bingefors[68]
Oat	Oat Land Race cv. Grey Winter	Goodey[73]
	Milford	McLeod *et al.*[74]; Anderson[75]
	Manod	Davies[76]
	Earley Grey Winter Unique, S 172	Howard and Cotten[72]
	Curt	Brendler *et al.*[77]
	Greta	Clamot[78]
	Algerian, Elan, Avon and Cassia	Stanton
Corn	Inrafruh, Inrakorn	Kuthe and Dern[79]
Garlic	Branco Mineiro, Cajuru do Reino, Mostarda, Sacaia de Goiania, Ihumas A, Centenario, Roxo Tardio, Campineiro, Roxo Perola Cacador and Barbado	Curi *et al*[80].
Rye	Heertvelder	Kort[81]
Wheat	Leone	Grujicic[82]

which have subsequently served as parental lines for breeding *D. dipsaci* resistant alfalfa cultivars.

Although host resistance provides an useful method for the management of the nematode, it is important to keep in mind that not all populations of nematode respond uniformly to a given resistant variety. For example, 'Wando' pea is susceptible to *D. dipsaci* population from Raleigh, North Carolina (RNC population), but resistant to Waynesville N.C. population[62].

Inheritance of resistance has been studied in few crops. Resistance in alfalfa is tetrasomic while in oats it is governed by a single factor pair and in clover by few genes.

Nematophagous fungi, *Arthrobotrys anconia, A. oligospora, Dactylella* sp. and *Catenaria anguillulae* offer some potential as biological agents for controlling the nematode. One to 5 conidia of *Hirsutella rossiliensis* kill *D. dipsaci* after 5 days by invading the body cavity[64]. In another study, two strains of *Monacrosporium salinum* have been recorded effective in trapping the nematode[65].

Ditylenchus angustus (Butler, 1913) Filipjev, 1936, The Rice Ufra Nematode

It is a parasite of rice plants and the disease caused by it is known as 'Ufra'. It is also known as the rice stem nematode. It was first reported in 1913[83] from areas now forming the teritory of Bangladesh. The disease was found to be quite prevalent in the areas of flooded paddy cultivation. At that time about 6 million acres of paddy fields were reported to be infected by the nematode.

Morphology

Compared to *D. dipsaci, D. angustus* is slightly slender nematode with low and flattened lip region. Tail possesses a sharply pointed tip.

Distribution

Bangladesh, India,, Vietnam.

Host Range

Rice.

Biology

The nematode follows an ectoparasitic mode of life and can be found at the peduncle base, the stem just above the upper nodes or inside the panicle glumes[84]. At harvest, the nematode present in the tissues, becomes inactive and in coiled stage. These cottony

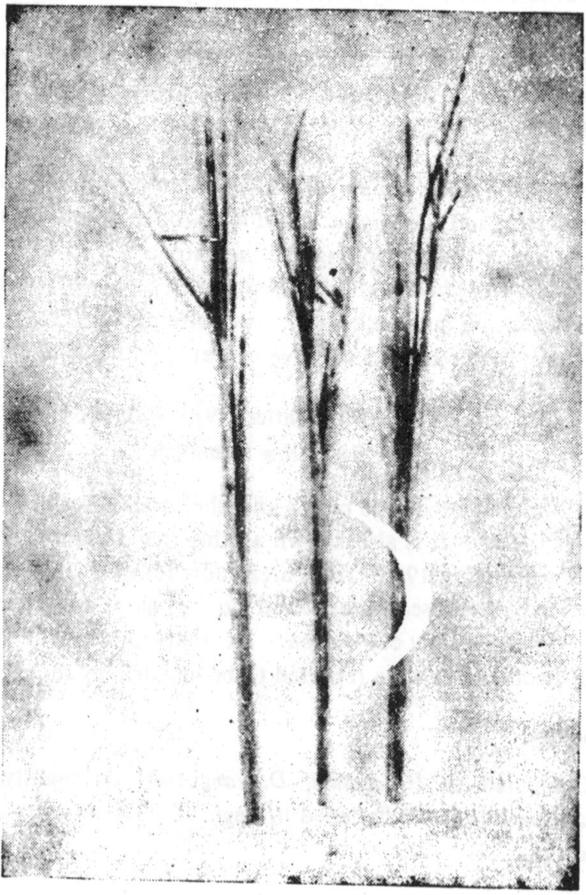

Fig. 5. *Ditylenchus angustus* damage to rice seedlings. Panicle enclosed within (left) sheath.

masses of quiescent nematodes remain in the stubble[85] and infect the crop in the next growing season. The nematodes climb up the stem, in humid conditions, and attack the stems, leaves and invade the growing point. The infloresence is invaded also which may result in failure in formation of grain heads or if formed, these may contain only few shrivelled grains. At the heading stage, nematodes collect between the inner side of leaf sheath and the developing ear for feeding. Branching of stem or even development of 3 to 4 distorted ears inside a single boot leaf sheath may occur also.

Fig. 6. *Ditylenchus angustus* on rice.
Panicle producing normal grains near tip only.

Fig. 7. *Ditylenchus angustus* on rice.
Distortion of panicles.
(After Hague, 1978).

Nematodes in the quiescent stage (wool stage) can remain in a desiccated condition for over 6 months and the spread is mainly through seedlings, irrigation water and farm implements.

Symptoms (Figs. 5-7)

The earliest symptom of damage in fields is chlorosis of upper leaves. Two distinct categories of symptoms, swollen 'ufra' and ripe 'ufra', have been recorded. In the former case, the panicle remains enclosed within the leaf sheath with the infected portions tending to branch. In the ripe 'ufra' case, the panicle emerges and produces normal grains only near the tip. In the nematode infested plants, the peduncle becomes dark brown with flowers, in the lower part, remaining unfertilized. In few cases, the panicle emerges partially[86].

Control

Destruction of infested stubbles followed by ploughings to keep fields dry for 2 to 3 months helps in destruction of the nematode. In the standing crop, spraying diazinon (100 ppm) copiously on the foliage and soil may help in reducing nematode populations. Application of phorate at 40 kg/ha to soil may also help in minimising plant damage[87].

Ditylenchus destructor Thorne, 1945, The Potato Rot Nematode

It is commonly known as the potato rot nematode because of the nature of damage to the potato tubers. It was first recorded in 1888[87] followed by reports from U.S.A. as well as some European countries. It is rarely observed to infest above ground plant parts, mainly confining itself to subterranean parts. It is quite a serious problem in some parts of Europe.

Morphology

Morphologically, these are similar to *D. dipsaci* except for a slightly robust body, lateral field marked with six incisures and basal bulb of the oesophagus generally [illegible] extended in a lobe reaching over the dorsal side of intestine. Populations [illegible] from different fields may exhibit quite a bit of variations which are probably [illegible] host induced.

Distribution

U.S.A., Europe, China.

Host

Besides potato, it attacks a wide variety of plants like sugar-beet[88], iris, hop, mint, tulips, red clover, etc.

Biology

The nematode can survive at temperature as low as —28°C. It can develop and multiply in 60-80 days at temperatures varying between 5-34°C. It is not very tolerant to drought conditions. It appears to thrive very well under U.S.S.R. conditions since about 10-20 per cent potato crop is reported to be damaged by this nematode.

The nematode multiplication is favoured by ammonium nitrate fertilization but amide forms of fertilizers depress reproduction[89]. The nematode can be cultured easily on fungi like *Alternaria tenuis, A. solani*, etc. and probably, in nature, survives by feeding on the fungi, if host plants are not available[90].

Symptoms

On potato, the infestation is characterised by production of tiny, white chalky spots (Fig. 8). These spots are quite superficial and do not extend deep into the tuber. The nematode may enter the tuber either in the region of eye or through lenticels. As a result of penetration, young developing and already differentiated cells of potato are disrupted. This type of reaction is at variance with the more specialised *D. dipsaci* which disrupts only differentiated cells with the plant responding back with de-differentiation of cells surrounding the nematode[91]. The tiny chalky white spots ultimately increase in size engulfing large portions of the tissue. The infested area appears granular with a honey combed structure. It is followed by a progressive drying and shrinking of skin, which becomes papery[92]. Apart from the direct effect of the nematode infestation on tubers, the lesions facilitate penetration of soft rot bacteria as well as saprophytic fungi resulting in rotting of tubers[93—95]. In such cases, a distinct area exists between the decayed portions and the uninvaded tissues and nematode populations are present, in abundance, in such areas.

On iris bulbs, the husk is shredded at the base, with the basal plate and the outer fleshy scale separated by a dark sunken crevice. Yellowish grey or black streaks are present in the fleshy scales and the leaves are distorted. The stem my exhibit light yellow or grey areas attached to the bulbs.

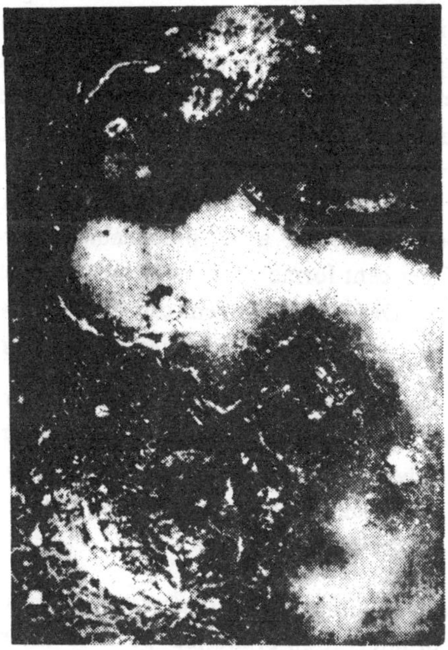

Fig. 8. *Ditylenchus destructor* damage to potato tuber. (After U.C., Riverside slide).

Symptoms, more or less similar to potato are present in sugar-beets. The nematodes enter the tissues near the crown. Small discoloured areas appear below the cortex that may engulf the entire upper part of beet. The infection by secondary micro-organisms results in rotting of tissues. On carrots, hops and *Ipomea batata* also similar damage symptoms are produced[90,96,97].

As a result of infection by the nematode on different hosts, there is a decrease in the carbohydrate content of the tissues with increased activities of peroxidase and polyphenol oxidase. Accumulation of polyphenol oxidase results in necrotic tissues[98]. Reduced amount of starch and increased amounts of amylase and amololytic activity have been recorded in infected potato tubers[99].

Control

Because of the polyphagous nature of the nematode and additionally because of its ability to feed on soil fungi, in absence of host, crop rotation is generally not applicable against this nematode. Soil fumigation by ethylene dibromide or granular chemicals like furadan or heterophos can reduce the populations effectively.[100,101] Heterophos granules applied before planting at the rates of 20 or 50 g/m^2 are reported to reduce infection as well as increase potato yields.[102] Hot water treatment of planting materials at 43°C for 10 hours, has been suggested as a practicable control measure.[103] Addition of 0.5 per cent formalin in water increases efficacy of hot water treatment. Dry heat treatment can be effective also as has been found in the case of iris bulbs treated at 5:°C for three hours.

In the recent past, attempts are now directed towards finding resistant sources. It has been suggested that forms immune to *D. destructor* can be obtained from *Solanum catarthrum, S. sucrense, S. vernei, S. famatinae, S. spegazzim, S. simplicifolicum, S. acaule, S. chacoense, S. demissum* and *S. jamesi*[104].

On the basis of different approaches employed in the control of the nematode, an integration of all the possible methods has been suggested and is claimed as a viable strategy for control of the nematode on potato. The steps suggested are :

(1) Use of rye or lucerne as the preceding crop ;

(2) Winter sowing of rye followed by late spring planting of potato ;

(3) Careful sorting of seed potatoes and early harvesting of potatoes ; and

(4) Soil treatment with metham sodium at 250 kg/ha[105].

Ditylenchus myceliophagus Goodey, 1958, The Mushroom Nematode

The species is of considerable economic importance in mushroom cultivation. Prior to 1958, the nematode encountered in mushrooms was recognised as *D. destructor*. Later it was found that it is an altogether different species and subsequently described as *D. myceliophagus*[106]. It is commonly encountered in all the mushroom cultivations and it has been estimated that populations as low as 3 nematodes/100 g compost, at spawning, can reduce yields by about 30 per cent. At a population level of 20, 100 and 300 nematodes/100 g at casing, there is a yield reduction of 50, 68 and 75 per cent respectively[107]. More than 20 nematodes/100 g compost can prevent cropping.

The nematode is similar to *D. destructor* but differs in certain morphological features like the length of post-vulval sac and length of the oesophageal overlap on the intestine. However, under periods of starvation, body length as well as other morphological measurements may get altered. The nematode is a migratory parasite and destroys the mycelium of the cultivated mushrooms. Temperature has an important bearing on the life cycle duration. It takes 40 days at 13°C and 21 days at 23°C[108,109]. At temperatures less than 13°C, there is very little development of the nematodes and it has been suggested that mushrooms can be successfully cropped under cool conditions, However, it does not appear to be an economically viable method.

As mushroom cultivation is of short duration and also is a perishable commodity, use of chemicals is generally not advocated for fear of residues. Biological control offers a possible method of reducing nematode populations. A predatory Hyphomycete, *Arthrobotrys robusta*, strain *antipolis* has been employed to reduce nematode populations. It can be grown alongwith the mushroom and has been recorded to reduce nematode populations by 40 per cent and increase mushroom yields by 20 per cent. The predatory fungal preparation is being marketted under the commercial name Royal 300[110-112].

REFERENCES

1. Thorne, G., Principles of Nematology. McGraw-Hill Book Company, Inc. N.Y., Toronto, London, 553 pp., 1961.

2. Kuhn, J., Uber des Vorkommen von *Anguillula* an erkrankten Bluhtenkopfen von *Dipsacus fullonum* L. *Zeitscher. wiss. Zool.*, 9, 129-127, 1857.

3. Steiner, G. and Scot, C.E., A nematosis of Amsinckia caused by a new variety of *Anguillulina dipsaci*. *J. Agr. Res.*, 49, 1087-1092, 1935.

4. Filipjev, I.N. and Schurmans Stekhoven, J.H., A manual of agricultural helminthology. Leiden, E.J. Brill, 878 pp.

5. Filipjev, I.N., On the classification of the Tylenchinae. Proc. *Helminthol. Soc. Wash.*, 3, 80-82, 1936.

6. Siddiqi, M.R., Tylenchida : Parasites of plant and insects. CAB Publication, U.K., 645 pp., 1986.

7. Yuksel, H.S.: Observations on the life cycle of *Ditylenchus dipsaci* on onion seedlings. *Nematologica*, 5, 289-296, 1960.

8. Goodey, T., On the susceptibility of clover and some other legumes to stem-diseases caused by the eelworm, *Tylenchus dipsuci*, syn. *devastatrix*, Kuhn. *J. Agr. Sci.*, 12, 20-30, 1922.

9. Barrett, J., Metabolic responses to anabiosis in the fourth-stage juveniles of *Ditylenchus dipsaci* (Nematoda). *Proc. R. Soc. Lond*, B., 216, 159-177, 1982.

10. Warton, D.A. and Barrett, J., Ultrastructural changes during recovery from anabiosis in the plant parasitic nematode, *Dilylenchus*. *Tissue and Cell*, 17, 79-96, 1985.

11. Sturhan, D., Biological races. pp. 51-71. In : Plant parasitic nematodes. Eds : Zuckerman, B., Mai, W.F. and Rohde, R.A., Vol. 2., Academic Press, 347 pp., 1971

12. Seinhorst, J.W., Some aspects of the biology and ecology of stem eelworm, *Ditylenchus dipasaci*. **Nematologica**, 2 (suppl.), 355-361.

13. Steiner, G., The problem of the taxon in the nematode genus *Dilylenchus* and its agricultural implications. *Proc. 14th Int. Congr. Zool.*, Copenhagen, 1953, pp. 377-379, 1956.

14. Eriksson, K.B., Crossing experiments with races of *Ditylenchus dipsaci* on callus tissue cultures. *Nematologica* 11, 244-248, 1965.

15. Webster, J.M., Significance of biological races of *Ditylenchus dipsaci* and their hybrids. *Ann. App. Biol.*, 59, 77-83, 1967.

16. Sturhan, D., *Mitt. Biol. Bundesanst.* Berlin-Deblem 136, 87-98, 1969.

17. Viglierchio, D.R., Race genesis in *Ditylenchus dipsaci. Nematologica*, 17, 386-392, 1971.

18. Chitwood, B.G., Haasis, F.A. and Blanton, F.S., Hot water-formalin treatment (at 110° to 111°F) of field-grown and forced narcissus bulbs infected with the bulb and or stem nematode, *Ditylenchus dipsaci. Proc. Helminthol. Soc. Wash.*, 8, 44-50, 1941.

19. Johnson, D.E. and Lear, B., Additional information regarding the hot water treatment of seed garlic cloves for the control of *Ditylenchus dipsaci. Plant Dis. Reptr.*, 49, 898-899, 1965.

20. Lear, B. and Johnson, D.E., Treatments for eradication of *Ditylenchus dipsaci* in clover and garlic. *Plant Dis. Reptr.*, 46, 635-639, 1962.

21. Winfield, A.L., Control of *Ditylenchus dipsaci* (Kuhn) Filipjev on narcissus in eastern England. Rep. 8, *Int. Symp. Plant Nematol.* (Antibes), 128, 1965.

22. Winfield, A.L., Factors affecting the control by hot water treatment of stem nematode *Ditylenchus dipsaci* (Kuhn) Filipjev in narcissus. *J. Hort. Sci.*, 45, 447-456, 1970.

23. Hwang, S.W. and Sayre, R.M., A method for the freezing and recovery of viable *Ditylenchus dipsaci* larvae. *J. Nematol.*, 1, 292-293, 1969.

24. Sayre, R.M. and Hwang, S.W., Freezing and storing *Ditylenchus dipsaci* in liquid nitrogen. *J. Nematol.*, 7, 199-202, 1975.

25. Faulkner, L.R., Bower, D.B., Evans, D.W. and Elgin, J.H. Jr., Mass culturing of *Ditylenchus dipsaci* to yield large quantities of inoculum. *J. Nematol.*, 6, 126-129, 1974.

26. Elgin, J.H., Peaden, R.N. and Faulkner, L.R., Evaluation of alfalfa for stem resistance. *Crop Sci.*, 15, 43-47, 1975.

27. Elgin, J.H. Jr., Peaden, R.N., Faulkner, L.R. and Evans., D.W., Effects of temperature and alfalfa entry on seedling reaction of stem nematode. *Crop Sci.*, 14, 669-672, 1974.

28. Griffin. G.D., The pathogenecity of *Ditylenchus dipsaci* to alfalfa and the relationship of temperature to plant infection and susceptibility. *Phytopathology*, 58, 929-932, 1968.

29. Grundbacher, F.J. and Stanford, E.H., Effect of temperature on resistance of alfalfa to the stem nematode (*Ditylenchus dipsaci*). *Phytopathology*, 52, 791-794, 1962.

30. Hanna, M.R. and Hawn, E.J., Seedling inoculation studies with the alfalfa stem nematode. *Can. J. Plant Sci.*, 45, 357-363, 1965.

31. Sayre, R.M. and Mountain, W.B., The bulb and stem nematode (*Ditylenchus dipsaci*) on onion in south eastern Ontario. *Phytopathology*, 52, 510-516, 1962.

32. Wallace, H.R., Movement of eelworms. 2. A comparative study of the movement in soil of *Heterodera schachtii* Schmidt and of *Ditylenchus dipsaci* (Kuhn) Filipjev. *Ann. Appl. Biol.*, 46, 84-94, 1958.

33. Seinhorst, J.W., Population studies on stem eelworms (*Ditylenchus dipsaci*). *Nematologica*, 1, 159-164, 1956.

34. Griffin, G.D., Effect of environment on association of *Ditylenchus dipsaci* to alfalfa. *Phytopathology*, 56, 879 (Abstr.), 1966.

35. Griffin, G.D., Hunt, O.J. and Murray, J.T., Pathogenecity of *Ditylenchus dipsaci* to sainfoin (*Onobrychis viciaefolia* Scop). *J. Nematol.*, 7, 93-94, 1975

36. Huisingh,, D. and Sherwood, R.J., The role of calcium in resistance of alfalfa to *Ditylenchus dipsaci.* Nematologica, 14, 8-9, 1968.

37. Sherwood, R.J. and Huisingh, D., Calcium nutrition and resistance of alfalfa to *Ditylenchus dipsaci. J. Nematol.*, 2, 316-323, 1970.

38. Fielding, M.J., Observations on the length of dormancy in certain plant infesting nematodes. *Proc. Helminthol. Soc. Wash.*, 18, 110-112, 1951.

39. Goodey, T., Quiescence and reviviscence in nematodes with particular reference to *Tylenchus tritici* and *T. dipsaci.*, J. Helminthol., 1, 47-52, 1923.

40. Rostrup, S., Oversight over landbrugsplanternes sygdomme 1 1912. *Tidsskr. Planteavl.*, 20, 275 pp., 1913.

41. Chitwood, B.G., Newhall, A.J. and Clement, R.L., Onion bloat or eelworm-rot, a disease by the bulb or stem nematode, *Ditylenchus dipsaci* (Kuhn) Filipjev. *Proc. Helminthol. Soc. Wash.* 7, 44-51, 1940.]

42. Caubel, G., Charpentier, F. and Guy, F., Comple renus de seances de l' Academia d' Agriculture de France, 71, 719-729, 1985.

43. Grill, D., The diseases and pests of theleek. Phytoma No. 371, 39-41, 1985.

44. Boelter, R.H., Gray, F.A. and Delaney, R.H., Effect of *Ditylenchus dipsaci* on alfalfa mortality, winter kill and yield. *J. Nematol.*, 17, 140-144, 1985.

45. Krusberg, L.R., Studies on the culturing and parasitism of plant parasitic nematodes, in particular *Ditylenchus dipsaci* and *Aphelenchoides ritzemabosi* on alfalfa tissues. *Nematologica*, 6, 181-200, 1961.

46. Bingefors, S., On resistance to stem nematode in alfalfa. *Nematologica*, 7, 17 (Abstr), 1962.

47. Blake, C.D., The etiology of tulip-root disease in susceptible and in resistant varieties of oats infested by the stem nematode, *Ditylenchus dipsaci* (Kuhn) Filipjev, 2. Histopathology of tulip-root and development of the nematode. *Ann. Appl. Biol.*, 50, 713-722, 1962.

48. Dijkstra, J., Symptoms of susceptibility and resistance in seedlings of red clover attacked by the stem nematode, *Ditylenchus dipsaci* (Kuhn) Filipjev. *Nematologica*, 2, 228-236, 1957.

49. Hussey, R.A. and Krusberg, L.R., Histopathology of resistant reaction in pea seedling to two populations of *Ditylenchus dipsaci Phytopathology*, 58, 1305-1310, 1968.

50. Hussey, R.A. and Krusberg, L.R., Histopathology of and oxidative enzyme patterns in Wando peas infected with two populations of *Ditylenchus dipsaci. Phytopathology*, 60, 1818-1825, 1970.

51. Chang, D.C.N., Campbell W.F. and Griffiin, G.D., Effects of temperature on fine-structural responses in the hypocotyl of alfalfa lines to *Ditylenchus dipsaci. J. Nematol.*, 7, 239-246, 1975.

52. Muse, B.D., Moore, L.D., Muse, R.R. and Williams, Pectolytic and cellulolytic enzymes of two populations of *Ditylenchus dipsaci* on 'wando peas' (*Pisum sativum L.*). *J. Nematol.*, 2, 118 124, 1970.

53. Krusberg, L.R., Hydrolytic and respiratory enzymes of species of *Ditylenchus* and *Pratylenchus. Phytopathology*, 50, 9-22, 1960.

54. Reidel, J.M. and Mai, W.F., A comparison of pectinases from *Ditylenchus dipsaci* and *Allium cepa* callus tissue. *J. Nematol.*, 3, 174-178, 1971.

55. Johnson. L.R., Observations on the occurrence of *Anguillulina dipsaci* (Kuhn) 1958 on rhubarb in Yorkshire. *J. Helminthol.*, 14, 77 84, 1936.

56. Hawn, E.J. and Hanna, M.R., Influence of stem nematode

infestation on bacterial wilt reaction and forage yield of alfalfa varieties. *Can. J. Plant Sci.*, 47, 203-208, 1967.

57. Caubel, G. and Samson, R., Effect of the stem nematode, *Ditylenchus dipsaci* on the development of 'cafe au lait' bacteriosis in garlic (*Allium sativum*) caused by *Pseudomonas flurescens*. *Agronomica*, 4, 311-313, 1984.

58. Windrich, W.A., Attraction of immature female nemas by male of the tulip race of *Ditylenchus dipsaci* in leaf tissue of tulip. *Nematologica*, 20, 570, 1974.

59. Caubel, G., Ducom, M. and Marre, R., Methyl bromide fumigation against *Ditylenchus* in seed and bulb lots. *Bull. OEPP*, 15, 17-22, 1985.

60. Russel, C.C., The stem nematode in south-eastern United States, p. 157-173. In : Nematology in the south region of the United States. *Soc. Coopt. Ser. Bull.*, 276, 206 pp., 1982.

61. Zinoveva, S.V., Kochurova, A.I. and Gerasimova, L.I., The use of potassium metabisulphite tablets for reducing the loss of onion and garlic caused by the stem nematode. Taksonomya i biologiya fitogel' mintov. Ed. Turlygina, E.S. Moscow, U.S.S.R. Nauka, pp. 62-70, 1984.

62. Eriksson, K.B., Nematode diseases of pasture legumes and turf grasses. pp. 66-96. In : Economic Nematology, Ed. Webster, J.M., Academic Press, N.Y., 1972.

63. Williams, A.S., Control of alfalfa stem nematode, *Ditylenchus dipsaci* (Kuhn) Filipjev by rotation, and additions to the host range. *Rep. Int. Symp. Plant Nematol.* (Antibes), 129, 1965.

64. Cayrol, J.C. and Frankowskl, J.P., Influence of the number of parasitizing condia of *Hirsutella rhossiliensis* on the mortality of *Ditylenchus dipsaci*. *Revue de Nematologie*, 9, 411-412.

65. Peloille, M., Frankowski, J.P. and Cayrol, J.C., Study of the four *Monacrosporium* strains in view of their use in biological control of phytophagous nematodes. *Acta Oecologica*, 5, 245-257, 1984

66. Hunt, C.J., Faulkner, L.R. and Peaden, R.N., Breeding for nematode resistance. In : C.H. Hanson (ed.) Alfalfa Science and Technology. Madison, American Society of Agronomy, Inc., 812 pp., 1972.

67. Moore, K.C. and McLeod, R.W., Stem nematode resistant lucerne could pay off. *Agric. Gaz.* N.S.W. 91(3), 28-30, 1980.

68. Bingefore, S., Resistance to nematodes and the possible value of induced mutations. In : Mutation Breeding for Disease Resistance. pp. 209-235. International Atomic Energy Agency, Vienna, 1971.

69. Brown, E.B., A broad red clover resistant to stem eelworm. *Pl. Path.* 8, 124, 1959.

70. Fiddian, W.E.H. and Aldrich, D.T.A., The susceptibility of red clover varieties to clover stem eelworm. *Pl. Path.*, 13, 139-143, 1964.

71. Bingefors, S., The use of nematode resistant varieties of grasses and legumes. *Herb. Abstr.* 39, 107-111, 1969.

72. Howard, H.W. and Cotton, J., 'Nematode resistant crop plants. In : Plant Nematology (J.F. Southey ed.), pp. 313-325. Her Majesty's Stationery Office, London, 1978.

73. Goodey, T., Observation on the susceptibility of certain varieties of oats to tulip-root caused by the stem eelworm, *Anguillulina dipsaci. J. Helminth.*, 15, 203-214, 1937.

74. McLeod, K., Golightly, W.H. and Price, C.D., Tulip root of oats in Durham. *Agriculture*, London, 61, 229-231, 1954.

75. Anderson, S., Angrepp av Stjalknematod (*Ditylenchus dipsaci*, (Kuhn) Filipjev) i havre. *Vaxtskyddsnotiser* 30, 10-12, 1966.

76. Davies, T.G., A spring oat variety resistant to the stem eelworm. *Nature*, London, 186, 813, 1960.

77. Brendler, R.A., Isom, W.H., Radewald, J.D. and Shibuya, F.,

Oat variety testing for tolerance to nematode causing 'tulip root'. *Calif. Agric.*, 25(7), 14-15, 1971.

78. Clamot, G., 'Greta' premiere variete belge d'avoine resistante de la tige. *Revue Agric.* Brux 21 ex., 1205-1211.

79. Kuthe, K. and Dern, R., Erfahr ungen bei der untersuchung von *Ditylenchus*—Be fall an Mais (*Zea mays*) in Hessen. *Gesunde* pft 22, 101-104, 1970.

80. Curi, S.M., Silveira, S.G.P., Siqueira, W.J. and Prates, H.S., Compontamento de cultivares de alho em relasao ao nematoide *Ditylenchus dipsaci. Nematologia Brasileira*, 9, 13, 1985.

81. Kort, J., Nematode diseases of cereals of temperate climates. In : J.M. Webster (Ed.) Economic Nematology, London and New York, Academic Press, 97-126, 1972.

82. Grujicic, G., Opojavi Stabjikine nematode (*Ditylenchus dipsaci*) (Kuhn) na pseniu u Jugoslaviji. *Savrener, Poljopr., Novi Sad* 10, 86-94, 1962.

83. Butler, E.J., An eelworm disease of rice. Agr. Res. Inst., Pusa, (India) Bull., 34, 1913.

84. Padwick, G.W., Manual of rice diseases. Commonw. Myc. Inst. Lond., pp. 104-115, 1950.

85. Taylor, A.L., In Tropical Nematology. Ed. J.E. Peachey, Tech. Commun. No. 40, Commonw. Bur. Helminth,. 264-268, 1969.

86. Cox, P.G., Symptoms of 'ufra' disease of deep water rice in Bangladesh. *Int. Rice Res. Newslett.*, 5, 8, 1980.

87. Haai, V.H., In Nematodes of Tropical Crops. Ed. Peachey, J.E. Commonw. Bur. Agric., London. 269-287, 1969.

88. Kuhn, J., New Erfahrungen auf dem Gebiete de Zuckerruben-kultur. *Jahrb. deut. Landwirts. Ges. fur* 1889, 4, 93-94, 1890.

89. Artem'ev, Y.M., *Dnylenchus destructor* on potatoes in the ural region and the effect of nitrogenous fertilizers on *D. destructor* infection of potatoes. Sbornik Mauchnykh. Trudov. Saratous-kogo Sel' skophozyatvennogo Instituta No. 54, 30-37, 1976.

90. Foot, M.A. and Wood, F.M., Potato rot nematode, *Ditylen-chus destructor* infecting hops in Newzealand. *J. Exptl. Agric.*, 10, 443-446, 1982.

91. Kostyuk, N.A., Interrelationships between some nematodes o the sub-family Anguininae and their plant hosts. *Trudy Gel. mentologicheskot. Laboratori*, 30, 30-45, 1980.

92. Thorne, G., *Ditylenchus destructor* n. sp., the potato rot nematode and *Ditylenchus dipsaci*, the teasel nematode. *Proc. Helminth. Soc. Wash.*, 12, 27-33, 1945.

93. Janowicz, K. and Mazurkiewicz, K., An attempt to evaluate effect of *Ditylenchus destructor* on infections of potato tubers with *Rhizoctonia solani. Ochrono Rosten*, 26, 8-9, 1982.

94. Janowicz, K., *Ditylenchus destructor* and fungi-causing dry rot in stored potatoes. *Ochrona Rosten* 28, 10-11, 1984.

95. Poloznenets, V.M., The influence of the intensity of stem nematode infestation of potato tubers on the development of bacterial diseases. *Khozyatstua*, 146-148, 1984.

96. Sten, M., Potato tuber eelworm, *Ditylenchus destructor* attack and damage to potato and carrot. *Gartneryrket*, 67, 1034-1036, 1977.

97. Ding, Z.F. and Lin, M.S., Identification of stem nematodes from sweet potatoes, potatoes and mint in China. *Acta Phytophylactica Sinica*, 9, 169-172, 1982.

98. Okophyi, N.S., Comparative study of physiological and biochemical mechanisms of the pathogenesis of *Ditylenchus* and *Globodera* infestations on potatoes. In : Steblevye Nematody sel' sknozyar stvennyph kul tur Mery b or bys Nimi. Voronezh, USSR, Vserassuskii N *11* Zashchity Rastenii, 90-98, 1983.

99. Alova, O.B., Study of the effect of *Ditylenchus destructor* on hydrolytic processes in potato tubers. Vesnik Khar' kovskogo Universitita No. 262, 78-80, 1984.

100. Darling, H.M., Adams, J. and Norgren, R.L., Field eradication of the potato rot nematode, *Ditylenchus destructor*, a 29 year history. *Plant Disease*, 67, 422-423, 1983.

101. Chukantseva, N.K., Some aspects of the study of the potato stem nematode in the Central Cherhozem zone of the RSFSR. In: Steblevye Nematody sel' sknozyar stvennyph kul'tur 1 Mery bor'bys Nimi. Voronezh, USSR, Vserassiskii N 11 Zashchity Rastenu, 11-27, 1983.

102. Vorona, V.F., Trials with heterophos against the potato stem nematode. Byulleten Vseroyuzhogo Instituta Gel'mintologu im K.Z. Skryabina No. 36, 69-70, 1983.

103. Shepshelev, Z.G., Erofeev, S.G. and Kovalevsku, U.V., Thermal treatment of tubers in the control of potato stem nematode. In : Zashchita Kartofelya ot poleznet i vreditcler, USSR. Nauchnolssledovatel'sku Instituut Kartofel'nogo Khozyatstva, 140-145, 1984.

104. Bukasov, S.M., Olefir, V.V. and Turuleva, L.M., Resistance of some species of wild potato to the stem nematode. In : Steblevye Nematody, sel' sknozyar-stvennyph Kul'tur i mery bor'bys nimi Voronezh, USSR. Vserosousku N 11 Zashchity Rastenu, 68-78, 1983.

105. Adylova, N.A. and Vasilevsku, V.N., The potato stem nematode at the collective farm 'Soph' in the fergara valley and its control. In : Steblevye Nematody sel'sknozyar-stvennyph kul'tur i mery bor'bys nimi Voronezh, USSR. Vserossusku N 11 Zashchitey Rastenie 59-61, 1981.

106. Goodey, J.B., *Ditylenchus myceliophagus* n. sp. *Nematologica*, 3, 91-96, 1958.

107. Arnold, N.P. and Blake, C.D., Some effects of the nematodes *Ditylenchus myceliophagus* and *Aphelenchoides composticola* on,

the yield of the cultivated mushroom. *Ann. Appl. Biol.*, 61, 161-166, 1968.

108. Cayrol, J.C., Importance des Maladies Vermiculaities dans les Champigemonnieres Francaises. *Mushr. Sci.*, 5, 480-496, 1962.

109. Evans, A.A.F. and Fisher, J.M., Development and structure of population of *Ditylenchus myceliophagus* as affected by temperature. *Nematologica*, 15, 394-402, 1969.

110. Cayrol, J.C., Biologische Bekampfung von Nematoden in Champignon bitneben mit Hilfe des Pitzes *Arthrobotrys robusta* var. *antipolis* (Royal 300). *Champignon*, 19, 30, 1978.

111. Cayrol, J.C., Frankowski, J.P., Laniece, A. and D'Hardenmare, G., Contre des nematodes en Champegneniere. *Revue Horticolè*, 184, 23-30, 1978.

112. Cayrol, J.C. and Frankowski, J.P., Lutte centre les nematodes mycophages parasites des d'un hyphomycete predateur du genre *Arthrobotrys*. *Mushroom Sci.*, 10, 407-413, 1979.

9

The Genus *Meloidogyne* Goeldi, 1892

The root-knot nematodes of the genus *Meloidogyne* are one of the most important plant parasitic nematodes with an extensive host range and wide distribution in temperate as well as tropical areas of the world. It is of historical as well as nematological interest since sustained nematological investigations all over the world have started with this organism. A brief account of its historical aspects is given here because of the importance of the nematode.

The root-knot disease was first reported in 1855, on the glass-house grown cucumbers in England by Berkeley who named the nematode as 'Vibrios'[1]. It was later described under the name of *Anguillula radicicola*[2]. At a later date, Cornu recorded root galls on sainfoin and as he was not aware of the first name, he gave the name *Anguillula marioni*[3]. Muller in 1884[4] observed some eggs within the females and described them as *Heterodera radicicola*. Goeldi in 1889[5] named the root-knot nematodes on coffee as *Meloidogyne exigua* (this paper was actually printed in 1892). Some other designations used during that period were : *Anguillula arenaria*[6], *Tylenchus arenarius*[7], *A. violae*, *Heterodera marioni*[8] and *Oxyuris incognita*[9]. Cobb in 1924[10] erected the genus or subgenus *Caconema* for the root-knot nematodes as related to *Heterodera*. Goodey[11] (1932), while reviewing the status of the root-knot nematodes, did not favour a separate genus for them and preferred to call them *Heterouera marioni*.

The nomenclatorial confusion continued till about the middle of the twentieth century when Chitwood[18] revived the genus *Meloidogyne*, originally erected by Goeldi in 1892 and recognised four existing species namely, *M. exigua* Goeldi, 1892 ; *M. javanica* Treub, 1885 ; *M. incognita* Kofoid and White, 1919 ; and *M. arenaria* Neal, 1889. A new species *M. hapla* and a new variety *M. incognita acrita* were added in the list. Subsequently, many new species were described. Presently, there are 61 species and two sub-species under this genus (Table 1). The most common species encountered throughout the world are *M. incognita*, *M. javanica*, *M. arenaria* and *M. hapla*.

TABLE 1

Presently known species of *Meloidogyne*

1. *Meloidogyne acronea* Coetzee, 1956
2. *M. africans* Whitehead, 1960
3. *M. aquatilis* Ebsary and Eveleigh, 1983
4. *M. ardenensis* Santos, 1968
5. *M. arenaria* (Neal, 1889) Chitwood, 1949
6. *M. artiellia* Franklin, 1961
7. *M. bauruensis* (Lordello, 1956) Esser, Perry and Taylor, 1976
8. *M. brevicauda* Loos, 1953
9. *M. camelliae* Golden, 1979
10. *M. caraganae* Shagalina, Ivanova and Krall, 1985
11. *M. carolinensis* Eisenback, 1982
12. *M. chitwoodi* Golden, O'Bannon, Santo and Finlay, 1980
13. *M. coffeicola* Lordello and Zamith, 1960
14. *M. cruciani* Garcia-Martinez, 1982
15. *M. decalineata* Whitehead, 1968
16. *M. diconincki* Elmiligy, 1968
17. *M. elegans* de Ponte, 1977
18. *M. enterolobii* Yang and Eisenback, 1983
19. *M. ethiopica* Whitehead, 1968

20. *M. exigua* Goeldi, 1892
21. *M. fujinensis* Pan, 1985
22. *M. grahami* Golden and Slana, 1978
23. *M. graminicola* Golden and Birchfield, 1965
24. *M. graminis* (Sledge and Golden, 1964) Whitehead, 1968
25. *M. hapla* Chitwood, 1949
26. *M. incognita* (Kofoid and White, 1919) Chitwood, 1949
27. *M. incognita acrita* Chitwood, 1949
28. *M. incognita wartellei* Golden and Birchfield, 1978
29. *M. indica* Whitehead, 1968
30. *M. inornata* Lordello, 1956
31. *M. javanica* (Treub, 1885) Chitwood, 1949
32. *M. kikuyensis* de Grisse, 1960
33. *M. kirjanovae* Terentyeva, 1965
34. *M. kralli* Jepson, 1983
35. *M. litoralis* Elmiligy, 1968
36. *M lordelloi* de Ponte, 1969
37. *M. lucknowica* Singh, 1969
38. *M. mali* Itoh, Ohshima and Ichinohe, 1969
39. *M. magadora* Whitehead, 1968
40. *M. Megatyla* Baldwin and Sasser, 1979
41. *M. megriensis* (Poghossian, 1971) Esser, Perry and Taylor, 1976
42. *M. microcephala* Cliff and Hirschmann, 1984
43. *M. microtyla* Mulvey, Townshend and Potter, 1975
44. *M. naasi* Franklin, 1965
45. *W. nataliei* Golden, Rose and Bird, 1981
46. *M. oryzae* Maas, Senders and Dede, 1978
47. *M. oteifae* Elmiligy, 1968
48. *M. ottersoni* (Thorne, 1969) Franklin, 1971
49. *M. ovalis* Riffle, 1963

50. *M. pini* Eisenback, Yang and Hartman, 1985
51. *M. platani* Hirschmann, 1982
52. *M. poghossianae* Kirjanova, 1963—species inquiernda
53. *M. propora* Spaull, 1977
54. *M. querciana* Golden, 1979
55. *M. salasi* Lopez, 1984
56. *M. sewelli* Mulvey and Anderson, 1980
57. *M. sinensis* Zhang, 1983
58. *M. spartinae* (Rau and Fassuliotis, 1965) Whitehead, 1968
59. *M. subarctica* Bernard, 1981
60. *M. suginanensis* Toida and Yaegushi, 1984
61. *M. tadshikistanica* Kirjanova and Ivanova, 1965
62. *M. thamesi* (Chitwood in Chitwood, Specht and Havis, 1952) Goodey, 1963
63. *M. turkestanica* Shagalina, Ivanova and Krall, 1985

Morphology

The species of *Meloidogyne* are usually differentiated on the basis of the morphological and morphometric characters of stylet, second-stage larvae, females and males[13]. The perineal pattern of the females (Fig. 1, 4) however, has been traditionally depended upon for identification of species. The four most common species, *viz.*, *M. incognita*, *M. javanica*, *M. arenaria* and *M. hapla* have been differentiated additionally on the basis of female, male and larval morpholgy[14].

In addition to conventional morphological characters, recent emphasis has shifted to the use of electrophoretic analysis of enzyme phenotype for the identification of species or biological races. Three of the four major species have been identified with a high degree of accuracy on their esterase phenotype (Table 2)[14]. For instance, all the populations of *M. javanica* are characterized in having three esterase bands which are unquestionably highly species-specific. *M. incognita*

TABLE 2

Enzyme phenotypes of the four common *Meloidogyne* species

Nematode species and populations	Number of populations	Enzyme phenotypes (Number following letter designation indicate major band of the phenotype)				Chromosome number
		Est*	Mdh*	Sod*	Got*	
(1)	(2)	(3)	(4)	(5)	(6)	(7)
M. arenaria (*A*)	7	A1	N1**	JA2	N1	36-52
	19	A2	N1	JA2	N1	37-54
	1	A2	N1	A4	N1	54
	12	A2	N3	A4	N1	54
	1	A2	N3	A4	H2	54
	1	A2	N3b	A4	N1	49
	4	A2	N3	JA2	N3	42-54
	1	A2	N1	JA2	N3	43
	7	A3	N1	A4	N1	52-54
	1	A3	N1	JA2	N1	52-54

(*Contd.*)

(Contd.)

(1)	(2)	(3)	(4)	(5)	(6)	(7)
M. arenaria (atypical populations)	2	**S1-M1	N1	I2	N1	37-43
	4	S1-M1	N3	I2	N1	36-42
	3	S2-M1	N1	I2	N1	36-42
	1	M3-F1	N1	JA2	N1	42 13-17(n)
M. hapla (*H*)	32	H1	H1	H1	H2	30-48
	1	Null	H1	H1	H2	17(n)
	1	A1	H1	H1	H2	14(n)
M. incognita (*I*)	104	I1	N1	I2	N1	32-46
	1	I1	N3b	I2	N1	42
	1	S1	N1	I2	N1	42-45
M. javanika (*J*)	44	J3	N1	JA2	N1	41-48
	2	J3	N3	JA2	N1	42-46

*Est—Esterase ; Mdh—Malate dehydrogenase ; Sod—Super oxide dimutase ; Got—Glutamate—Oxalo acetate transaminase ;

**N—Nonspecific ; S, M, F—slow, medium, fast migration of band.

exhibits a distinct phenotype designated as I_1. Likewise, all the populations of *M. hapla* have esterase phenotype designated as H_1. The fourth major species, *M. arenaria* is not associated with any particular esterase phenotype (Table 2, Fig. 2)[14]. Malate dehydrogenase and glutamate oxaloacetate transaminase phenotypes are also useful in separating *M. hapla* from other major species.

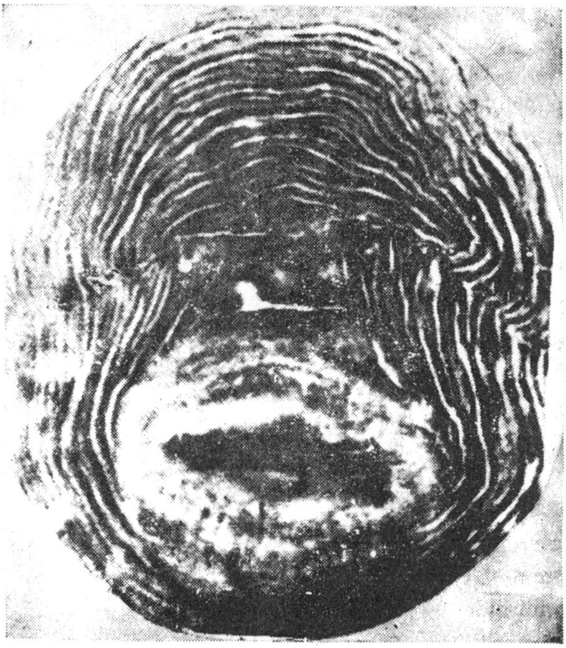

Fig. 1. Perineal Pattern of female of root-knot nematode.

There is a strong sexual dimorphism exhibited by the members of the genus. The males are vermiform, generally 1400 × 30μm, while the adult females are pear shaped (pyriform) and measure about 700 × 400μm. The outermost layer of the body wall is formed by non-cellular, elastic, basically triple layered cuticle, which is secreted by the underlying hypodermis. It maintains the body shape against body turgor, helps in locomotion and protects the internal systems from physical, chemical and biological forces. Externally, the cuticle bears several markings which are helpful in toxonomy.

In the female, the transverse annules are present in the anterior region, become shallow and sparse and gradually disappear in the anterior middle region of the body. These reappear in the posterior portion forming finger print like perineal pattern around the tail terminus, anus and vulva. There patterns are typical of each species. Remnants of lateral fields may be visible interrupting the transverse annules in the anterior and posterior regions.

The openings of the phasmids in juveniles and females are usually hidden by the annulations or aerolations in the lateral field. In females, the phasmids are visible between tail tip and anus in the perineal pattern, while in males the slit like phasmidial openings are located within the inner ridge of the lateral field, at the level of cloacal openings.

The oesophagus is typically tylenchoid with procorpus, valvated metacorpus, isthmus and three well developed oesophageal glands, usually extending ventrally. The duct from the dorsal gland joins the lumen of the procorpus close behind the knobs while those from subventral glands join in the metacorpus. The intestine joins the isthmus where a pair of specialized oesophago-intestinal cells (cardia) are present.

The preparasitic larva has a small genital primordium comprised of two small flat somatic cells and two large spherical germinal cells. The primordium enlarges after feeding begins and also during moults, finally giving rise to a pair of prodelphic ovaries in females and a single testis in male. Two anteriorly directed testes may be present in some males formed after sex reversal under environmental stress. Most species of *Meloidogyne* are able to reproduce parthenogenetically as well.

The eggs are about 40 μm in diameter and 90-100 μm long, ovoid bodies. The egg shell is triple layered, outer vitelline, middle chiitinous and inner glycolipid, which are secreted by the different parts of the uterus. The eggs are highly resistant to chemicals and environmental stress and have special significance in survival and persistence of the population.

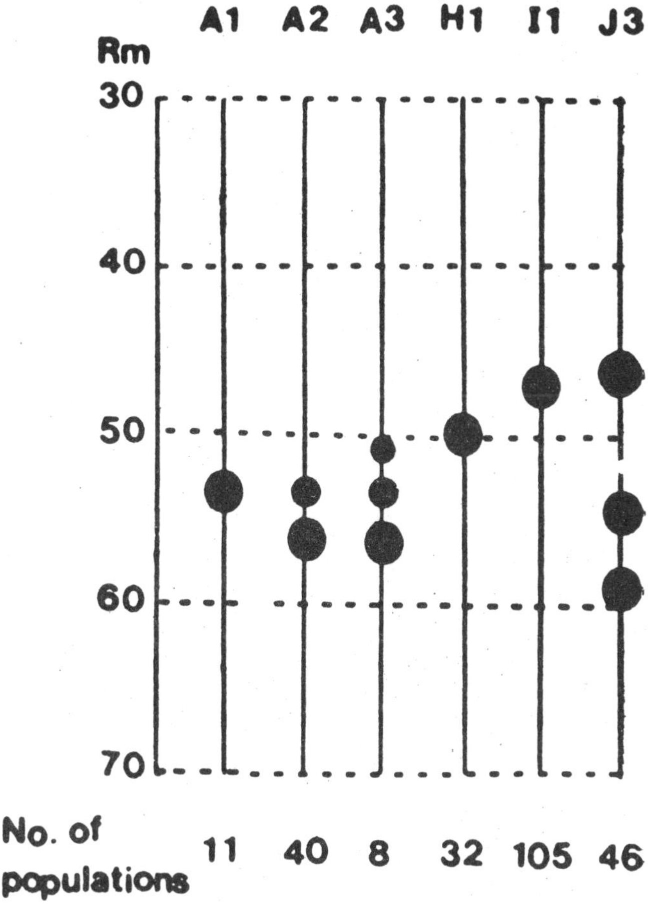

Fig. 2. Esterase phenotype of *Meloidogyne* species.
A1, A2, A3--*M. arenaria*;
H1—*M. hapla*; I1—*M. incognita*;
J3—*M. javanica*.
(Reproduced after Esbenshade and Triantaphyllou, 1985)

Biology

The eggs are laid in a gelatinous matrix secreted by the female at the posterior end The posterior part of the females may protrude out of the host tissue or may remain embedded in it. The contents of the egg undergo a series of determinate cleavages giving rise to the

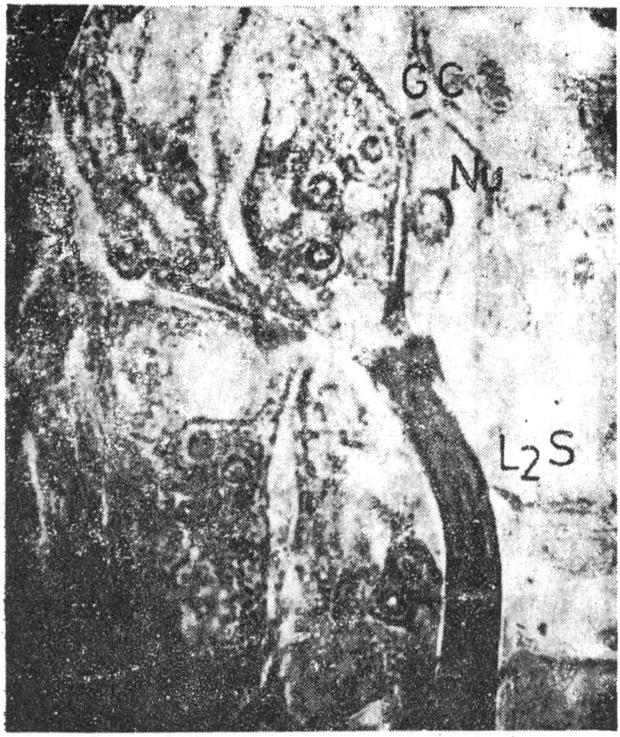

Fig. 3. Second-stage larva feeding around group of giant cells.

blastula and triploblastic gastrula stages. Finally, the first stage larva is formed. The embryo and first-stage larvae are highly resistant to moisture stress[15]. The first moult occurs within the egg to give rise to the second-stage larva that is ready to hatch out. The hatching can occur freely in water or moist soil without any requirement of root exudates or hatching factors. However, under adverse conditions of moisture, osmotic pressure, temperature, etc., the hatching is inhibited or postponed to ensure longer survival. The gelatinous egg matrix also assists in survival. For instance, due to hygroscopic nature, it absorbs ambient moisture and the moisture content in the egg sac remains higher than that in drying soil[16]. Further, under high degree of moisture stress the high osmotic pressure in the matrix may inhibit hatching.

The pre-parastic second-stage larva moves freely in soil in

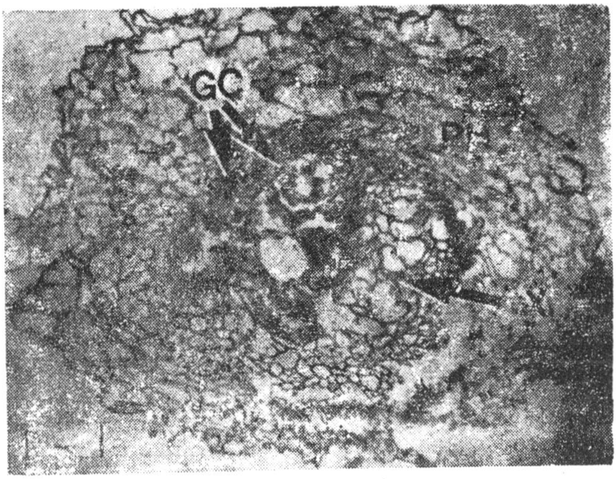

Fig. 4. Transverse section of root showing multinucleate condition of giant cells.

search of a suitable host tissue. The larva has sufficient glycogen and glycolipid reserved for survival for several months in case the host is not available. These are also capable of surviving under adverse conditions.

The larvae feed on the epidermal cells of the underground plant parts and penetrate the newly formed tissues above the meristematic zone. They usually orient parallel to stele, with head pointed opposite to the tip. The nematode may be found also attached to the more mature roots, since the root apex grows further.

The larva starts feeding usually on the pericycle cells. The cell contents are liquified and semi digested extracorporeally with the help of hydrolytic enzymes, secreted by oesophageal gland, eventually resulting in a group of multinucleate giant cells around the nematode head (Figs. 3 & 4). The cortical parenchymatous cells, around the giant cell, undergo excessive multiplication (hyperplasia) giving rise to tiny swellings (primary galls) on the roots, several of which merge to form big multiple galls.

The post-infectional second-stage larva continues to feed for a period of several weeks, gradually swelling and losing the capacity to move. The period of feeding (2-3 weeks optimally) varies with the host reaction, temperature, moisture and environmental factors. A series of three moults occur in quick succession. The cuticles of second, third and fourth stages are retained till the last moult. Hence, these stages have the projecting tail cuticle (spike) of the second stage. In most species, a majority of larvae develop into pyriform sedentary females with few monorchic males. Under environmental extremities, proportionately more males are produced and some of the larvae destined to be females actually produce diorchic males.

The reproduction may be without fertilization (mitotic parthenogenesis) or by cross fertilization (amphimixis). Still other species may reproduce through meiotic parthenogenesis. Cytogenetical informations related to some root-knot nematode species are presented in Table 3. The female lays about 400-500 eggs into a gelatinous matrix secreted by the rectal glands. The total number and size of eggs are affected by the host status and the level of environmental stress. As many as 2,882 eggs from a single female have been recorded. The egg laying is completed in a period of about a week. The eggs hatch freely in water but the rate of hatching may be higher if the host root exudates are present[17].

Two types of egg masses, white and brown, are formed depending upon environmental stresses. White egg masses are formed during early stage of crop growth and the eggs hatch immediately. Towards the end of growing season and/or under adverse environmental conditions, brown egg masses are formed which exhibit dormancy. This ensures survival of the nematodes under unfavourable conditions. Dormancy (diapause) of egg masses is also known in *M. naasi* where hatching is stimulated by chilling at 0 to 10°C for a week prior to incubation at 28°C. Dormancy has been recorded also in some races of *M. incognita*[15].

The total duration of life cycle, under optimum conditions, is

TABLE 3

Summary of cytogenetical informations related to Root-Knot Nematodes[18]

Meloidogyne species*	Populations studied (number)	Countries of origin (number)	Chromosome number n	2n	Mode of reproduction
(1)	(2)	(3)	(4)	(5)	(6)
M. carolinensis	2	1	18		
M. megatyla	1	1	18		Amphimixsis
M. microtyla	2	1	18-19		
M. subarctica	1	1	18		
M. exigua	6	1	18		
M. graminicola	1	1	18		
M. graminis	10	5	18		
M. ottersoni	1	1	18		Meiotic
M. naasi	1	1	18		
M. chitwoodi	6	3	14-18		
M. hapla (Race A)	48	24	13-17		Parthenogenesis
(Polyploid)	1	1	28		
(Polyploid)	2	2	34		
M. arenaria	18	13		30-38	
	34	21		40-48	
	68	32		51-56	
M. cruciani	1	1		42-44	
M. enterolobii	1	1		46	

(*Contd.*)

(*Contd.*)

(1)	(2)	(3)	(4)	(5)	(6)
M. hapla	6	3		30-32	
	11	8		43-48	Obligatory
M. hispanica	4	4		33-36	
M. incognita	6	6		32-38	
	215	64		41-46	Mitotic
M. javanica	126	45		42-48	
M. microcephala	3	2		36-38	
M. oryzae	2	1		51-55	Partheno-genesis
M. platani	1	1		42-44	
M. querciana	1	1		30-32	
M. spp. (from coffee)	6	3		50-56	
(from rice)	2	2		36	
Unidentified	5	3		36	
	11	7		40-46	
Total	602				

3-4 weeks in most species. However, the duration is affected by a number of ecological factors ,especially host status, temperature, moisture, pH, osmotic pressure, etc. It has been recorded that the duration from infective larva to egg laying female is least (17 days) at optimum temperature of 27.5-30°C but more (upto 57 or more days) at low and high temperatures.

Temperature, moisture and soil types are perhaps the most important abiotic factors influencing the distribution, survival, growth, reproduction and pathogenecity of root-knot nematodes. *M. incognita*, *M. javanica* and *M. arenaria* are essentially warm weather species. These species are known to occur in the areas having annual average temperature between 15-33°C. In contrast, *M. hapla* inhabits the area with annual average temperature less than 15°C[16,19]. The optimum temperature for embryogenesis is between 25-30°C for a number of species. In *M. javanica*, at 25°C, the hatching is 2 and 3 times more than at 20 and 15°C, respectively. In *M. chitwoodi* and *M. hapla*, hatching at 25°C is only about 2 times as rapid as at 15°C.

Meloidogyne species can be divided into two distinct survival groups, thermophylis and cryophylis, on the basis of their ability to survive lipid-phase transitions that occur at 10°C[15]. *M. hapla*, *M. chitwoodi* and *M. naasi* are [cryophylis and can survive at soil temperatures upto or below 0°C. At soil temperatures of 0-5°C, the larvae of *M. hapla* can survive and remain infective for 2-4 weeks. *M. incognita*, *M. arenaria* and *M. exigua* are thermophylic and do not have extended survival in soil below 10°C. Within a species, there may be temperature minimums between populations, indicating the existence of thermotypes. Generally, the optimum temperature for survival of eggs and larvae is from 10-15°C.

Movement of larvae is also governed by temperature. For example, at 14°C, 17 per cent of *M. hapla* larvae are able to migrate 20 cm and infect tomato roots in 10 days, while only 3 per cent of *M. incognita* at the same temperature can move and reach the host roots. The sequential order for cold resistance in some of the species is :

M. chitwoodi>*M. hapla*>*M. incognita*>*M. arenaria*>*M. javanica*[15].

Root-knot nematodes are active with soil moisture levels of 40-60 per cent field capacity. In sandy and clay soils, the optimum emergence of larvae from egg masses is slightly above the field capacity (pF 3.0). The gelatinous matrix, under drying conditions, offers protection against loss of water. Embryonic as well as first larval stages are more resistant than unhatched second-stage larvae within the egg shell. Under field conditions, the second-stage larvae can enter into a state of anhydrobiosis and prolong their survival in dry soils[15].

Generally, these nematodes occur most frequently in soils with less than 10 per cent clay, less than 30 per cent silt and atleast 60 per cent sand[20]. The mobilily of infective larvae and root invasion decreases with an increase in silt and clay fractions over sand and no migration occurs in soils having more than 30 per cent clay plus silt. Movement of second-stage larvae and root invasion are maximum when the ratio of soil particle diameter to length of second stage larvae is 1 : 3.

The nematode activity increases with an increase in oxygen concentration. A moist soil, which drains rapidly and allows oxygen concentration above 10 per cent, provides favourable environment for the development, hatching and mobility of root-knot nematodes[15]. A concentration of CO_2 above 10 per cent affect adversely activity of *M. hapla*. The survival, hatching and reproduction are generally not affected within pH range of 4.0-8.0 Maximum hatching in *M. javanica* has been recorded at pH 6.4-7.0 and reduced hatching at osmotic potential above 2.5 atmospheres.

The fate of second-stage larvae for further development depends on their ability to find host. It is now presumed that the infective larvae do not find hosts at random, but are attracted to plants in response to stimuli emanating from the roots. There are specific receptor sites located on the surface of second-stage larvae (cuticular receptor site) which serve for attachment of chemotactic factors or attracting molecules from plant roots. The other membrane receptor (sensory receptor) are located within the cephalic sensilli. Molecules of cuticular receptor sites which play an important role in recognition of host are located in that part of cuticular surface close to sensory organs. The binding of attractant molecules with specific to carbohydrate molecules, localized in the receptor sites at cephalic area of nematode cuticle, leads to host recognition. The specific carbohydrate moiety that occurs in cuticular receptor of *M. incognita* is characterized as mannose residues, which has been suggested to be derived as a secretion of cephalic sensilla[21,22].

The second-stage larvae can be attracted ; and migrate to host roots, 25 cm vertically within 10 days[23]. They accummulate in the region of cell elongation, just behind the root cap. They are also attracted to apical meristem or points where roots emerge or penetration sites of other invading nematodes or cut surfaces of roots[24].

The nature of stimuli emanating from root-rhizosphere and mechanism involved in the perception of these stimuli by the larvae are not clearly understood. Many organic and inorganic compounds including amino acids, sugars and CO_2 are considered to be important stimuli provided by roots and micro-organisms around them. It is possible that penetration may be accomplished also enzymatically through cellulase and pectinases.

Host Range

Collectively, root-knot nematodes are considered among the top five major plant pathogens. One or more of the species attack nearly every crop responsible for the world's supply of food, fibre, timber, resins, ornamentals or other cash crops[25]. More than 200 plant species are reported to be parasitized by root-knot nematodes.

M. incognita and *M. javanica* have the widest host ranges, covering over 232 and 144 genera of plants, respectively. The more preferred hosts are dicot vegetables, pulses, fibre, fruit and cash crops. Some monocots like sugarcane, pearl millet, banana, etc., are also heavily attacked. These species are often serious problems in nurseries of fruit trees/vines, vegetables and ornamentals. Certain crops like wheat, though normally non-hosts, can be attacked in heavily infested soil[26].

M. arenaria is a serious problem on groundnut, tomato, beans, eggplant, okra, tobacco, etc. *M. hapla* has a limited distribution in the cooler regions and the host range covers over 9 genera of plants. The most important host is groundnut. *M. graminicola* is known to attack plants belonging to over 21 genera, but is mainly a problem of rice growing areas.

Although the four major species are polyphagous, the other species appear to be rather host specific. For example, *M. exigua* has been recorded only from coffee ; *M. megatyla* from pine ; *M. brevicauda* from tea ; *M. mali* from apple ; *M. ovalis* from sugar maple ; *M. ardensis* from periwinkle ; *M. deconincki* from ash and *M. litoralis* from *Ligustrum* sp.

A world-wide survey, conducted under the International *Meloidogyne* Project and subsequent tests on differential hosts has indicated existence of races in two species of root-knot nematodes. Four races of *M. incognita* and two of *M. arenaria* have been established on the basis of the North Carolina host tests[20] (Table 4). The races differentiated on the basis of host preferences are morphologically indistinguishable and apparently unrelated to cytological races based on chromosome counts[20] (Table 5).

Distribution

Root-knot nematodes are known to occur throughout the world. Their occurrence has been reported from all the continents including Asia, Europe, North and South America and Australia. Four common species, *M. incognita*, *M. javanica*, *M. hapla* and *M. arenaria*, have been recorded in many areas of tropical and temperate regions. However, *M. hapla* and *M. naasi* are essentially confined to cooler areas.

Symptoms

Basically, these are parasites of roots or underground stem, pods, etc. The above ground symptoms, therefore, are of slow debility of root in its function of nutrient and water uptake and translocation. The plants may be dwarfed, yellowish with smaller foliage and poor and fewer fruits. The very broad leaved plants like tobacco or cucurbits may show day time wilting. The symptoms are often mistaken for macro or micro nutrient deficiency or moisture stress. Poor emergence and death of young seedlings may occur in heavily infested soil (Fig. 5) but death of grown up plants is rare unless some fungus (like *Fusarium*, *Rhizoctonia*, *Phytophthora* or *Verticillium* species) or bacteria are involved in a complex. In some instances, early but poor fruiting for a relatively short bearing period may be seen.

The below ground symptoms on the roots are small galls which, in case of multiple infection on the nearby tissue, may coalesce to form large galls (Figs. 6-7). Some species, like *M. hapla* induce small (primary) galls with branching of root above it while others

TABLE 4

North Carolina Differential Host Test Reaction Chart

Meloidogyne species and Race	Differential Host Plants					
	Cotton 'Deltapine 61'	Tobacco 'NC 95'	Pepper 'California Wonder'	Watermelon 'Charleston Grey'	Peanut 'Florunner'	Tomato 'Rutgers'
M. incognita						
Race 1	–	–	+	+	–	+
Race 2	–	+	+	+	–	+
Race 3	+	–	+	+	–	+
Race 4	+	+	+	+	–	+
M. javanica	–	+	–	+	–	+
M. arenaria						
Race 1	–	+	+	+	+	+
Race 2	–	+	–	+	–	+
M. hapla	–	+	+	–	+	+

TABLE 5

Chromosome Numbers and Modes of Reproduction of the Four Major Meloidogyne Species

Meloidogyne species and cytological race	Range of chromosome numbers	Mode of reproduction
M. incognita		
Race A	40-46	mitotic parthenogenesis
Race B	32-36	mitotic parthenogenesis
M. javanica		
M. arenaria		
Race A	54(50-56)	mitotic parthenogenesis
Race B	34-37	mitotic parthenogenesis
M. hapla		
Race A	14-17	facultative meiotic parthenogenesis
Race B	30-32, 43, 45, 48	mitotic parthenogenesis

like *M. incognita*, *M. javanica*, etc., may produce larger galls. The size of the gall also depends upon the characteristics of the plant, since the gall is the product of the reaction of the host to the parasite. For instance, the cucurbits develop much larger galls whereas chillies may have much smaller ones with the same species. Besides galling, some other symptoms may be forking of the tap root as in carrots or beet ; curly root tip or pimple like tubercles on tubers (*e.g.*, in potato) or pods (*e.g.*, in groundnut).

The general appearance of the crop in a field is of patchy growth due to uneven distribution of the nematode (Fig. 5). The poorly growing patches grow in size each year and spread more in the direction of plough and irrigation water. The symptoms

Fig. 5. Sugarbeet field infested with root-knot nematode exhibiting poor emergence and patchy growth.

advance with the age of the crop and are more spectacular under drought or low fertility levels. Usually, the injury is not detectable from the above ground symptoms until the crop has already suffered almost 25 per cent damage. Both, quality and quantity of produce are affected.

The symptomatolgy of root-knot nematode infested host plants may be summarised as :

(1) suppressed shoot-root ratio ;

(2) nutritional deficiencies in foliage showing chlorosis ;

(3) temporary wilting during periods of mild water stress or during mid-day when temperatures are high, even though adequate soil moisture is available ; and

(4) reduced crop yield[24].

Crop Loss

The root-knot nematode is often the only or one of the few nematode diseases of crops known to the farmer owing to the spectacular symptoms. One or more of the presently known species or sub-species are serious problem on almost all the plants of economic importance. Despite the importance of this group of nematodes, there are no accurate data on the enormity of crop losses because of their infestation. But, on the basis of general literary survey and experience, the average yield losses in the world are believed to be about 5 per cent. It is presumed to be much larger in the developing countries of the sub-tropics and tropics[24].

Besides the direct damage caused to the plant, the root-knot nematode are notorious for the disease complexes involving fungi, bacteria, viruses, mycoplasma, insects and other nematodes. The most important of these are the complexes formed by the root-knot nematodes and the root invading fungi causing wilt and root-rot diseases. The first report of a nematode-fungus complex was by Atkinson in 1892[27] involving the root-knot nematode and *Fusarium* wilt on cotton.

The nematodes are known to provide entry avenues and

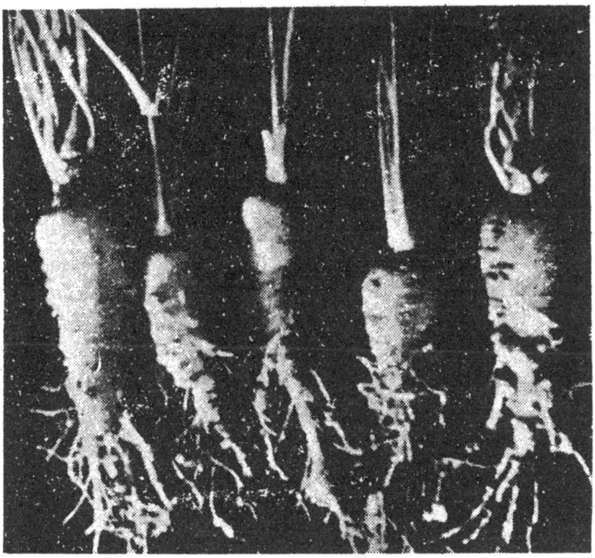

(a)

(b)

Fig.6. Root-knot nematodeg alls on) *a*) Carrot, (*b*) Black pepper.

(a)

(b)

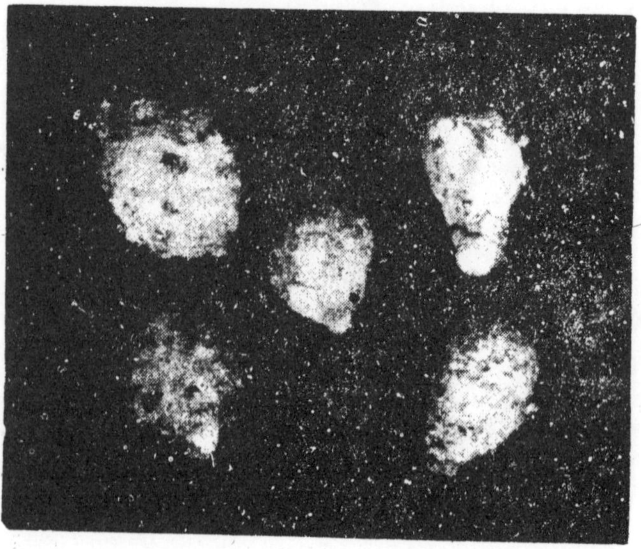

Fig. 7. **Root-knot nematode galls on** (*a*) Groundnut, (*b*) Potato.

modify the mineral composition and physiology of the host favouring fungal growth. The fungi, in turn, may help in the establishment of nematodes. Plugging of xylem vessels by wilt fungi interferes with mineral and water uptake and makes the plant more vulnerable to the nematode[28]. Contrary to this, secretions from fungal hyphae may modify and eventually kill the giant cells. This factor might be responsible for reduced population growth of *Meloidogyne*. The developing nematodes also produces or induces the plant to produce a factor transmitted upwards which breaks the resistance of the plant to fungus[29].

Several fungi, like species of *Rhizoctonia*, *Sclerotium*, *Fusarium*, *Phomopsis*, *Aspergillus*, *Corticium*, *Curvularia*, *Pythium*, *Verticillium*, *Phytophthora*, etc., have been known in association with the root-knot nematodes. The damage to the plant is often synergistic. The wilt causing bacteria like *Pseudomonas solanacearum*, *P. carryophylii* and *Corynebacterium michiganense* are known to damage the plant synergistically when root-knot nematode is present[30,31].

A number of mosaic and leaf curl viruses as well as mycoplasma like organisms are known to produce symptoms earlier in the presence of root-knot nematodes. Synergistic losses to mungbean have been reported in case of the insect, stem borer (*Stomoptera nertaria*) in combination with *M. javanica*.

The root-knot nematodes have also interesting relationships with other concomitant nematodes[32]. The relationship is usually neutral or stimulatory at low populations and antagonistic at high population densities. It depends upon the feeding site and mechanism. The competition is generally weak. The endoparasites, especially the sedentary ones, are more competitive than the ectoparasites. The root-knot nematodes are, therefore. the toughest competitors but in some cases may modify the host physiology to favour other nematodes like *Pratylenchus* spp.

Meloidogyne spp. are unique in their capability to induce giant cells as well as causing extensive pericycle hyperplasia and cortical hypertrophy resulting in galls[33]. The second stage infective larvae travel to the cortex intercellularly and establish feeding sites, which consist of a group of four to six multinucleate enlarged cells (giant cells) in the central vascular region. In soybean roots, it has been observed that there is an increase in the oxyreductase enzyme along the migration path of the larvae, suggesting increased metabolic activity in these cells[34]. This relationship is so balanced that continued existence of both the parasite as well as the giant cells is necessary for the development and maintenance of two antagonistic entities. Giant cells induced by these nematodes characteristically have transfer cell-like wall ingrowths, jutting into cytoplasm, and these are localized in specific regions of the giant cell wall which are adjacent to vascular tissues[35,36].

The events involved in the induction of giant cell by root-knot nematodes have not been as yet resolved conclusively. According to one view, several adjacent members of a row of undifferentiated cells, in central vascular region, coalesce through dissolution of their walls followed by combination of cell wall breakdown and repeated mitosis without cytokinesis, which ultimately lead to the condition of multinucleated giant cell[37]. According to the second view, the basic mechanism of development of giant cell involves only repeated mitosis without cytokinesis[38]. Apparently, this hypothesis overrules dissolution of the cell wall as the process of formation of giant cell. Galls usually develope 1-2 days after larval penetration in roots.

Hydrolysis of host proteins by nematode proteases may lead to release of amino acids including tryptophan. Synthesis of IAA from tryptophan and accumulation of the growth regulator at the site of growing root tip can induce gall formation. Furthermore, the bound IAA from host also may be released under the action of hydrolases emanating from the invading larvae.

Generally, the nematode infection also influences adversely the respiration, photosynthesis, water uptake and nutritional status of host and eventually leading to suppressed plant growth and yield[24]. In tomato, under experimental conditions, *M. javanica* is reported to cause reduced rate of photosynthesis, two days after inoculations. It also interferes with the production and translocation of root-derived factors regulating photosynthesis.

The females require considerable amount of energy for their survival and production of large number of eggs which they derive from host nutrition. In the infected plants, the nematodes serve as metabolic sink.[40,41]. Giant cells are considered as the intense site of metabolic activities which stimulate mobilization of photosyntbates from shoot to root.

Nutritional Imbalance

The nutritional status of root-knot nematode infected hosts is rather inconclusive and often contradictory. In general, impaired absorption and translocation of minerals and blocking metabolites at the site of infection have been recorded in many reports. Growth reduction, chlorosis and alteration of N, P, K, Ca, Mg and Fe concentration of tomato foliage are known. However, increased quantitiy of N, P, K and Mg in infected roots of tomato, as observed by many researchers, suggest impaired translocation from roots to foliage and mobilization of nutrients from shoot to roots[42-44].

Control

A number of methods like crop rotation, flooding, hot water, steam treatment, predaceous nematodes, chemicals and breeding resistant varieties have been experimented with, since long in Europe and America.

The hot water treatment of infested planting material such as root, tubers, etc., has been found effective. Treatment of infected papaya seedling roots at 50°C for 10 minutes and incubation of potato tubers at 45°C for 48 hrs have given satisfactory control provided the main field was not infested. Similar treatments may be effective in case of rooted cuttings of perennial vines, trees and ornamentals.

Use of dry or preferably moist heat to raise temperature of soil, sufficient to kill the nematodes, has proved an effective measure, especially where small amounts of soil are involved, *e.g.*, in nursery beds, pots, etc. Burning of crop waste or other matter over soil surface may give partial sterilization. Exposure of soil to solar heat by spreading on concrete floor or ploughing of field in summer can provide reasonably good control ; and efficiency can be improved by polythene mulching.

Among the cultural methods, adoption of land preparation methods causing turning and drying of soil, crop rotation, fallow, use of organic manures and amendments, rabbing, change of date of sowing, etc., have been most extensively experimented.

The root-knot nematodes survive in the form of eggs and second stage larvae in soil. Most of the eggs hatch in moist soil. The population of larvae declines by over 75 per cent in a period of 3-4 months due to starvation, desiccation and heat, etc., if weeds or other hosts are not available. Hence, one or two seasons without host are usually sufficient to bring down the population densities to enable a successful subsequent crop. Rotation with graminaceous poor hosts, such as wheat, barley, *rabi* maize, sorghum, etc., or dicots like mustard, sesame, *Crotalaria spectabilis*, asparagus, African marigold, etc., which also have some antagonistic effect, for one to two years are known to suppress root-knot nematodes. The very wide host range of these nematodes, both among cultivated and weed plants, is a great limitation necessitating thorough weed control to ensure the efficiency of rotation.

The problem of root-knot disease is increased especially in intensive cropping systems in sub-trophical climate where the vegetable and fruit grower, with limited land holdings, prefers to grow several successive crops within a year without leaving sufficient time between two crops to let the population decline. In such cases, the only alternative is to find out commercially profitable non-host crops or to resort to short rotations coupled with resistant varieties and chemical method.

Two year intensive cropping systems including sweet corn-soybean-wheat-soybean-spinach, and turnip-peanut-cucumber-turnip-cucumber-soybean have been found effective in suppressing *M. incognita* race-1 in USA[45]. Incorporation of mustard, sesame, maize, wheat, etc., suppress nematode populations in India. The potassic fertilizers are known to reduce root-knot intensity. Cultural practices like long periods of fallow or flooding are not practicable in agro-economic systems where the root-knot is a problem. Use of organic manures is of great value since the decomposition products as well as promotion of natural enemies decrease nematode populations.

Replacing susceptible varieties with genetically resistant ones appears to be the most plausible approach to nematode management. Unfortunately, polyphagous and polyspecific nature of nematode populations, existence and evolution of physiological races under selection pressure, difficulties of incorporating resistance into commercially preferred varieties, etc., are some of the chief problems encountered in a disease resistance programme. Currently, breeding for host resistance against root-knot nematodes is judged to be the most important method over other components of management. Sources of resistance have been located in a number of plant species against these nematodes (Table 6).

The role of enzymes such as peroxidase, IAA-oxidase superoxidedismulase, ribonuclease, phenylalanine ammonialyase, and phytoalexins in the mechanism of resistance have been examined by many workers[47].

Recent studies have also explored mechanism of plant nemtode interactions leading to resistance expression at molecular level. Incompatible reaction involving tomato—*M. incognita* and cowpea—*M. incognita* combinations has been shown to be under the influence of *de novo* gene activation at transcriptional level, resulting in the synthesis of new species of poly (A), containing mRNA and enzyme protein possibly required for the production of low molecular metabolits having antibiotic properties[48–51].

Considerable attention has been paid to evolve biological means to combat the root-knot and other nematodes. Major emphasis has been on nematophagous fungi, bacteria, predatory nematodes and antagonistic plants. Extracts of a large number of cultivated and weed plants have been found toxic to nematodes and the chopped parts of such plants when added to soil have resulted in reduced intensity of root-knot. Parts of *neem* (*Azadirachta indica*), *Melia azadirach*, *Zinnia*, marigold and several other plants have been found effective. Amendment of soil with non-edible oilseed cake of *neem*, *karanj*, *mahua*, etc., also have been found effective.

Recent experiments with a fungus, *Paecilomyces lilacinus*, which parasitizes eggs, have shown good promise[52]. A few beneficial associations between mycorrhizal fungi and root-knot nematodes

TABLE 6

Sources of resistance against Root-Knot Nematodes

Crop	Source of resistance species/lines/varieties	*Meloidogyne* spp.	Some resistant cultivars
(1)	(2)	(3)	(4)
Tamato (*Lycopersicon esculentum*)	*L. peruvianum*	*M. incognita*	
	L. pimpinellifolium	*M. javanica* *M. arenaria*	
Sweet potato (*Ipomoea batatus*)	Jenseyhypes	*M. incognita*	Heart Gold
	Tinian (PI 153655)	*M. javanica*	Nemagold
Bean (*Phaseolus vulgaris*)	Albama No. 1 Albama No. 2 PI 165426	*M. incognita*	Manca wonder

	PI 165435 B 4175		
Lima bean (*Phaseolus lunatus*)	Hopi Indians Hopi 5989 Nemagreen	*M. incognita*	Westan
Cucurbits	*Cucumis anguria*	*M. incognita*	
Cucumis spp.		*M. hapla* *M. javanica*	
Cucumis spp.	*C. ficifolius*		
Cucurbita spp.	*C. longipes* *C. metuliferus*	*M. incognita*	
Corn (*Zea mays*)	Pioneer 309	*M. incognita*	
Pepper (*Capsicum* spp.)	*C. frutescens*	*M. incognita*	Nemaheart Mississippi 68
Egg plant (*Solanum melongena*)	*S. sisymbriifolium*	*Meloidogyne* sp.	
Cowpea (*Yigna unguiculata*)	Iron	*M. incognita*	Missippi Silver

(*Contd.*)

(Contd.)

(1)	(2)	(3)	(4)
	Brabham	*M. javanica*	
	Catjan	*M. arenaria*	
Soybean (*Glycine max*)	Biloxi	*M. incognita*	Bragg
			Delmer
	Loredo		Centennial
Cotton (*Gossypium* spp.)	Clevewilt 6	*M. incognita*	NC 95
	Auburn 56		
Tobacco (*Nicotiana tabacum*)	Ti 706	*M. incognita*	

have also been recognized. The ectomycorrhiza, if established before nematode invasion, form a tight network of hyphae and modify the cortical tissue, thus forming the so-called Hartwig Net[53]. This provides a mechanical and physiological barrier to the invading nematodes. Some vesicular-arbuscular mycorrhizae like *Glomus* spp. are known to decrease the severity of root-knot[54] or increase plant tolerance to nematode by improving host nutrition and growth[55].

The chemicals used in early trials to control the root-knot nematodes included the halogenated hydrocarbon fumigants like DD, EDB, etc. Methyl bromide is being used still for sterilizing soil for experimental purposes and to a limited extent in tobacco cultivation[56]. The emulsifiable formulation of DBCP has been used commercially in the grapevines, tobacco, etc., but has been given up now due to the possible hazards of bromine residues. Certain other chemicals tried against these nematodes are the organophosphates and carbamates which are preferred because of the low dose (1.0 to 4.0 kg a.i./ha), granular formation and their systemic action. The OP compounds like fensulfothion, phosphamidon, dichlorofenthion, fenamiphos, chloropyriphos, ethoprophos and carbamates like aldicarb, carbofuran, dimethioate, methomyl, thiodemeton, etc., have been used also. However, applications in the main field have not been economic for most annual crops, despite invariably increased yields. High cost of the chemicals and their non availability have been the greatest limitations.

Alternative methods have been evolved to reduce the amount of chemical required. Limiting the application to nursery beds in transplanted crops ; bare root dip treatments and seed dressing with systemic nematicides have been the most promising approaches adopted by a large number of workers. These treatments protect the crop from nematode attack in the early stage.

Some analogues of carbofuran, *e.g.*, carbosulfan, UC-51229, etc., have shown basipetal systemic properties in reducing root-knot, lesion and other nematodes when applied as foliar sprays in tomato maize, potato, tobacco, peas, etc.

Certain other chemicals like amino acids, growth regulators, fatty acids, fungicides, antibiotics, homeopathic drugs, etc., have been used but with limited success. Recently, the filtrates of a number of fungi like species of *Fusarium*, *R'izoctonia*, etc., have shown nematicidal properties. Some compounds like avermectins produced by bacteria (*Streptomyces avermitilis*) have been found also effective against the root-knot nematodes.

Since the individual methods of nematode control have proved inadequate, attempts have been made to manage the root-knot and other nematodes through integrated systems combining several methods[67].

The use of nematicides can be limited to nursery bed in transplanted crops or can be used as seed dressers in transplanted as well as direct seeded crops. Bare root dips as well as spot or row applications in main field, in respect of perennial and other high value cash crops, can reduce the dosages of chemicals. Additionally, these treatments can be combined with summer ploughing, organic manuring, supplementary fertilizers, proper irrigation, resistant varieties, whenever available, weed control and crop rotation with commercial graminaceous or other non-host crops or antagonistic cover crops.

REFERENCES

1. Berkeley, M.J., Vibrio forming excrescenes on the roots of cucumber plants. *Gard. Chron.*, 14, 220, 1855.

2. Greef, R., Ueber nematoden in Wartzelenschwellungen (Gallen) verschiendener Pflanzen. Sitzber, Ges. Natur. W. Merburg, 169-174, 1872.

3. Cornu, M., E'tudes Sur le *Phylloxera vastatrix. Mem. Press. acad. Sci.* Paris, 26, 164-174, 1879.

4. Muller, C., Mittellungen uber die unseren Kulturpflanzen Schadlichen das Gaschlecht *Heterodera* bildenden Wurmer. *Longwirtz. Jahrb.*, 13, 1-42, 1884.

5. Goeldi, E.A., Relatoria Sobre a molestia do Caferiro na provincia do Rio de Janeiro. *Arch. Mus. Nac., Rio de Janeiro*, 8, 7-121, 1892.

6. Neal, J.C., The root-knot disease of peach, orange and other plants in Florida due to work of *Anguillula. U.S. Dept. Agr. Div. Ent. Bull.*, 20, 1889.

7. Cobb, N.A., *Tylenchus* and root gall. *Agr. Gaz. N.S. Wales*, 1, 213-214, 1890.

8. Marcinowski, Kati, Parasitizche und semiparasitizche an pflanzenlebenden Nematoden. *Arb. Kaiserlichen biol. Anstalt Land- u. Forst. Berlin*, 7, 1-192, 1909.

9. Kofoid, C.A. and White, W.A., A new nematode infection of man. *J. Amer. Med. Assoc.*, 72, 567-569, 1919.

10. Cobb, N.A., Notes on nemas. *Proc. Helminthol. Soc., Wash. J. Parasitology*, 202, 103, 1924.

11. Goodey, T., On the nomenclature of root-gall nematodes. *J. Helminthol.*, 9, 21-28, 1932.

12. Chitwood, B.G., Root-knot nematodes Part I. Revision of the genus *Meloidogyne* Goeldi, 1887. *Proc. Helminthol. Soc. Wash.*, 16, 90-104, 1949.

13. Siddiqi, M.R., Tylenchida : Parasites of Plants and Insects. CAB Publications, U.K., 645 pp., 1986.

14. Esbenshade, P.R. and Triantaphyllou, A.C., Identification of major *Meloidogyne* species employing enzyme phenotypes and differentiating characters, 135-142. In : An : An Advanced Treatise on *Meloidogyne*, vol. 1, Biology and Control., Eds. Sasser, J.N. and Carter, C.C., 135-142, 1985.

15. Van Gundy, S.D., Ecology of *Meloidogyne* spp.—Emphasis on environmental factors affecting survival and pathogenicity. In : An Advanced Treatise on *Meloidogyne*, vol. 1, Biology and Control., Eds. Sasser, J.N. and Carter, C.C., 177-182, 1985.

16. Wallace, H.R., The influence of soil moisture on survival and hatch of *Meloidogyne javanica*. *Nematologica*, 14, 231-242, 1968.

17. Viglierchio, D.R. and Lownsbery, B.F., The hatching response of *Meloidogyne* species to the emanations from the roots of germinating tomatoes. *Nematologica*, 5, 153-157, 160.

18. Triantaphyllou, A.C., Cytogenetics, cytotaxonomy and phylogeny of root-knot nematodes. In : An Advanced Treatise on *Meloidogyne*, vol. 1, Biology and Control, Eds. Sasser, J.N. and Carter, C.C., 113-126, 1985.

19. Sasser J.N. and Carter, C.C., Root-knot nematode (*Meloidogyne* spp.) identification, morphological and physiological variations, host range, ecology and control. In : Nematology in the Southern Region of the United States. *So. Coopt. Ser. Bull.*, 276, 21-32, 1982.

20. Sasser, J.N. and Carter, C.C., Overview of the International *Meloidogyne* Project 1975-1984. In : An Advanced Treatise on *Meloidogyne*, vol. 1, Biology and Control, Eds. Sasser, J.N. and Carter, C.C., 19-24, 1985.

21. Zuckerman, B.M., Hypothesis and possibilities of intervention in chemoresponses. *J. Nematol.*, 15, 173-182, 1983.

22. Zuckerman, B.M. and Jansson, H.B., Nematode chemotoxis and possible mechanism of host/prey recognition. *Ann. Rev. Phytopath.*, 22, 95-113, 1984.

23. Prot, J.C., Vertical migration of four natural populations of *Meloidogyne*. *Rev. Nematol.*, 1, 21-26, 1978.

24. Hussey, R.S., Host parasite relationships and associated physiological changes. In : An Advanced Treatise on *Meloidogyne*, vol. 1, Biology and Control, Eds. Sasser, J.N. and Carter, C.C., 143-154, 1985.

25. Sasser, J.N., Pathogenecity, host range and variability in *Meloidogyne* species. In : Root-knot Nematodes (*Meloidogyne* species) : Systematic Biology and Control, Eds. Lamberti, F.J. and Taylor, C.E., 257-268, 1979.

26. Patel, G.J., Patel, D.J., Jogani, D.K. and Patel, S.T., Wheat—a new host of *Meloidogyne javanica* in India. *Indian J. Nematol.*, 16, 34, 1986.

27. Atkinson, G.F., Some diseases of cotton. *Ala. Agric. Expt. Stn. Bull.* (Auburn) 41, 65, 1892.

28. Webster, J.M., Interaction of *Meloidogyne* with fungi on crop plants. In : An advanced Treatise on *Meloidogyne*, vol. 1, Biology and Control, Eds. Sasser, J.N. and Carter, C.C., 183-192, 1985.

29. Fattah, F.A. and Webster, J.M., Ultrastructural changes caused by *Fusarium oxyporium* f. sp. *lycopersici* in *Meloidogyne javanica* induced giant cells in *Fusarium* resistant and susceptible tomato cultivar. *J. Nematol.*, 15, 128-135, 1983.

30. Moura De, R.M., Echandi, E. and Powell, N.T., Interaction of *Corynebacterium michiganensis* and *Meloidogyne incognita* on tomato. *Phytopathology*, 65, 1332-1335, 1975.

31. Eisenback, J.D., Interactions among concomitant population of nematodes. In : An Advanced Treatise on *Meloidogyne*, vol. 1, Biology and Control, Eds. Sasser, J.N. and Carter, C.C., 193-206. 1985.

32. Johnson, A.W. and Nusbaum, C.J., Interaction between *Meloidogyne incognita*, *M. hapla* and *Pratylenchus brachyurus* in tobacco. *J. Nematol.*, 334-340, 1970.

33. Dropkin, V.H., Cellular responses of plants to nematode infection. *Ann. Rev. Phytopath.*, 7, 101-121, 1969.

34. Endo, B.Y. and Veech, J.A., The histochemical localization of oxidoreductase enzymes of soybeans infected with root-knot nematode, *Meloidogyne incognita*. *Phytopathology*, 59, 418-425, 1969.

35. Jones, M.G.K. and Northcote, D.H., Multinucleate transfer cells induced in coleus roots by the root-knot nematode, *Meloidogyne arenaria*. *Protoplasma*, 75, 381-395, 1972.

36. Jones, M.G.K. and Dropkin, V.H., Scanning electron microscopy of nematode induced giant transfer cells. *Cytobios*, 15, 149-161, 1976

37. Dropkin, V.H. and Nelson, P.E., The histopathology of root-knot nematode infections in soybean. *Phytopathology*, 50, 442-447, 1960.

38. Huang, C.S., and Maggenti, A.R., Wall modifications in developing giant cells of *Vicia faba* and *Cucumis sativus* induced by root-knot nematode, *Meloidogyne javanica. Phytopathology*, 59, 931-937, 1969.

39. Loveys, R.R. and Bird, A.F., The influence of nematodes on photosynthesis in tomato plants. *Physiol. Plant Path.*, 3, 5-529, 1973.

40. Bird, A.F. and Loveys, R.R., The incorporation of photosynthates by *Meloidogyne javanica. J. Nematol.*, 7, 111-113, 1975.

41. McClure, M.A., *Meloidogyne incognita* : a metabolic sink. *J. Nematol.*, 9, 88-90, 1977.

42. Dropkin, V.H. and King, R.C., Studies on plant parasitic nematodes homogenously labeled with radio phosphorus. *Expt. Parasitol.*, 5, 469-480, 1956.

43. Hunter, D.H., Nutrient absorption and translocation of phosphorus as influenced by the root-knot nematode (*Meloidogyne incognita acrita*). *Soil Sci.*, 86, 245-250, 1956.

44. Dasgupta, D.R. and Deb, D.L., Studies on mineral absorption and translocation in tomato plants inoculated with root-knot nematodes, *Meloidogyne javanica* and *M. arenaria*, separately and together with special reference to 32p. *Phytopath. Z.*, 75, 74-81, 1972.

45. Johnson, A.W., Specific crop rotation effects combined with cultural practices and nematicides. In : An Advanced Treatise on *Meloidogyne*, vol. 1, Biology and Control, Eds. Sasser, J.N. and Carter, C.C., 283-301, 1985.

46. Fassuliotis, G., Plant resistance to root-knot nematodes. In : Nematology in the Southern Region of the United States. *So. Coopt. Ser. Bull.*, 276, 33-49, 1982.

47. Huang, J.S., Mechanism of resistance to root-knot nematodes. In : An Advanced Treatise on *Meloidogyne* vol. 1, Biology and Control, Eds. Sasser, J.N. and Carter, C.C, 165-176. 1985.

48. Premachandran, D. and Dasgupta, D.R., A theoretical model for plant nematode interaction. *Rev. Nematol.*, 6, 311-314, 1983.

49. Parthasarathy, S. and Dasgupta, D.R., Changes in 32p-labelled RNA fractions from tomato roots inoculated with root-knot nematode, *Meloidogyne incognita. J. Nuclear Agri. Biol.*, 13, 10-15, 1984.

50. Raja, Aditi and Dasgupta, D.R., Enhanced synthesis of messenger RNA in relation to resistance expression in cowpea (*Vigna unguiculata*) infected with the root-knot nematode, *Meloidogyne incognita. Rev. Nematol.*, 9, 34-48, 1986.

51. Ganguly, A.K. and Dasgupta, D.R., Sequential development of peroxidase (E.C. 1.11.1.7) and IAA oxidase in relation to resistant and susceptible responses in tomatoes to the root-knot nematode, *Meloidogyne incognita. Indian J. Nematol.*, 9, 143-151, 1979.

52. Jatala, P., Biological control of nematodes. In : An Advanced Treatise on *Meloidogyne*, vol. 1, Biology and Control, Eds. Sasser, J.N. and Carter, C.C., 303-308, 1985.

53. Marx, D.H., Ectomycorrhizae as biological deterrents to pathogenic root infections. *Ann. Rev. Phytopathol.*, 10, 429-454, 1972.

54. Kellam, M.K. and Schenck, N.C., The effect of vesicular-arbuscular mycorrhizal fungus, *Glomus macrocarpus*, on the amount of galling produced by *Meloidogyne incognita* on soybean. *Proc. American Phytopathol. Soc.*, 4, 124, 1977.

55 Thomson Casen, K.M., Hussey, R.S. and Rancadori, R.W., Interaction of vesicular-arbuscular mycorrhizal fungi and phosphorus with *Meloidogyne incognita* on tomato. *J. Nematol*, 15. 410-417. 1983.

56. Hussaini, S.S., Efficacy of durofume, rabbing and aldicarb against root-knot nematode and weeds in tobacco nurseries. *Indian J. Nematol.*, 15, 88-92, 1985.

57. Gaur, H.S. and Prasad, S.K., An integrated approach to nematode population management towards efficient use of nematicides. Proc. National Conf. on Short and Long Term Hazards of Pesticides and Strategies for their Safe Use. INSA, Feb. 24-26, India, 1986.

10

Genera Heterodera Schmidt, 1871 and Globodera Skarbilovich, 1959

The cyst-forming nematodes of the superfamily Heteroderoidea are one of the most important pests of agricultural crops. Earlier, these were considered as problems in the temperate regions only but during the last two decades or so, several species, of economic importance, have been reported from the tropical as well as sub-tropical areas too. The record of new species or already known species from these areas is more because of intensified research activities rather than the belief of unfavourable climatic conditions for these nematodes in the tropical or sub-tropical regions.

The first cyst-forming nematode was discovered in 1859 in Germany, where it was found associated with sugarbeets exhibiting a sickness referred to as rubenmudigkeit or beet tiredness[1]. The nematode was subsequently named as *Heterodera schachtii* in 1871[2]. Later, cyst nematodes were described infesting pea and potato and were named respectively as *H. goettingiana* and *H. rostochienisis*[3,4]. However, these names did not gain acceptance and all the species were grouped together as *H. schachtii* and the cysts on different crops were identified only as the race or variety of the species *schachtii*. The specific status of the different forms, recognised under the name of *H. schachtii*, were reviewed in 1940[5] which led to recognition of five valid species : *H. schachtii* (type species), *H. goettingiana*, *H. rostochiensis* (now *Globodera rostochiensis*), *H. major* (now *H. avenae*) and *H. punctata* (now *Punctodera punctata*). Till

...ut 1970, all the cyst-forming nematodes were placed in the genus *Heterodera*, under the family Heteroderidae and superfamily Tylenchoidea. A new taxon in the superfamily rank was proposed in 1971[6], raising it to a superfamily rank, Heteroderoidea, incorporating two families, Heteroderidae and Nacobbidae. A new cyst-forming genus, *Sarisodera* was proposed in 1971[7] which differed from *Heterodera* on the basis of sunken vulva and no fenestration. Subsequently, five additional genera (*Bidera* Krall and Krall, 1978[8]; *Cactodera* Krall and Krall, 1978[8]; *Dolichodera* Mulvey and Ebsary, 1980[9]; *Globodera* Behrens, 1975[10]; and *Punctodera* Mulvey and Stone, 1976[11]) were proposed on the basis of differences in the cyst morphology. The erection of the genera, *Bidera* and *Cactodera* has not been accepted generally since it is felt that the genus *Heterodera* is a complex one requiring considerable additional investigations before any generic realignments could be proposed.

Apart from morphological features, host status of a population has been also considered important in the identification of the species of this group. In fact, this was one of the important criteria employed in naming a new species. However, the value of cone top structures in the identification of the species was demonstrated in 1954[12] and subsequently in 1955[13] definitions of the various parts of the cone top structures were put forward which have proved to be great help in the identification of the species.

Morphology

Most of the economically important species belong either to *Heterodera* or *Globodera*. Presently, there are 56 species in the genus *Heterodera* and 12 in *Globodera*. All the species are root parasites. There is strong sexual dimorphism in the group with males remaining as vermiform and free living for most of the part of their adult life as against the adult females which become swollen into flask-shaped or sub spherical immobile form. After the death of the female, the body wall turns into tough, resistant brown cyst, containing embryonated eggs. Ultimately, these cysts get detached from the roots and remain dormant in soil till suitable environmental conditions are available for emergence of larvae.

Members of this group are highly adapted towards plant parasitism. The female cuticle is strongly annulated and is usually modified in the posterior part of the body into lace-like pattern. The adult females are sedentary with a short neck. Vulva is equatorial to usually terminal and tail is absent. Males are vermiform with strong annulation and sclerotisation in the cephalic region. Testis is single and outstretched. Tail is bluntly rounded or hemispherical. Bursa is absent.

Distribution

As stated already, for quite some time, these nematodes were recognised as problems of temperate regions only. This contention was based mainly on the fact that, in the past, most of the recorded species were from temperate regions only. During the past decade or two, there has been an intensification of nematological activities all over the world, especially in the African, Middle East and Asian countries, with the result that a large number of nematode species, hitherto recorded from temperate regions only, have been recorded from these countries as well. A glance through the literature shows that of the total number of species recorded within the genera *Heterodera* and *Globodera*, a little over one third of them have been recorded from tropical and sub tropical areas. Apparently, the record of new species or already known species is more because of intensified activities on research on plant parasitic nematodes rather than the belief of unfavourable climatic conditions for cyst species in these regions. However, most of the species, recorded so far, thrive well under cool conditions. The distribution of the various species, is related also to the cultivation of the host crops besides other influencing factors like soil type and structure, rainfall, etc.

Dissemination

As is true with other plant parasitic nematodes, lateral spread, by their own movement in a field is slow. Apart from the farm implements, transport of infested soil through human and animal feet, transport of infested plant material from one area to another, dust and sand storms and overflow of water through infested areas also play a major role in disseminating nematode cysts from one place to another[14].

Biology

The eggs and larvae present inside the cysts or also in egg-sacs in some species, are the basic units which initiate the life processes. A detailed account of the developmental stages in the egg, before the second-stage larvae emerge from the egg shell has been given for *H. schachtii* in 1950[15]. The second-stage larvae are the infective units, which under favourable conditions of temperature and moisture are released through the hatching pores (fenestra) of the cyst and move in soil till host roots are encountered. Generally, the larvae penetrate the roots behind the growing tip or at the point of initiation of lateral root. They migrate through the cortex, inter- as well as intracellularly, disrupting the cortical cells in the process, and reach the stele region where they take a position parallel to the root axis. The anterior region of the larvae is positioned in the cells of the pericycle while the tail region is towards the root tip.

Emergence of larvae, from the cysts, may continue intermittently for several days, with most of them coming out in the first flush. The process of root penetration also, therefore, may continue for several days. Most of the root penetration, however, is completed within a fortnight period. The development of the second-stage, inside the root, is completed within the next 7-8 days depending on the prevailing temperature conditions. There are two further moults before they become adults. The genital primordium becomes visible in the third-stage larvae and it also shows a well developed rectum. At this stage, the female larvae exhibit paired ovaries and the males a single testis. As the development proceeds, the females become flask shaped while the males remain more or less cylindroid. In some species, a gelatinous matrix is exuded at the posterior end of the adult female, representing an egg-sac and which may contain few to several eggs. In *H. cajani*, the egg sac is often twice as big as the female. The adult male emerges from the root and becomes free living in soil. Since it does not feed, it lives long enough to fertilize the protruding females before it dies.

Emergence of Larvae from Cysts

A period of 'dormancy' is reported for several species. Generally, this period coincides with the end of the crop season. Various factors like age of cysts, temperature to which they are exposed[16], effect of host on maturity of cysts[17], presence of hatch inhibitor in old cysts[18], etc , have been suggested as factors influencing 'dormancy' of cysts However, with the availability of appropriate host and suitable temperature conditions, the larvae within the cysts get activated and are released in soil. For some species, host root diffusate is essential for good emergence. In some cases, like *H. schachtii*, root diffusates from a non-host crop (*Hesperis matronalis*) can activate also emergence of larvae from the cysts. The extent to which larvae will emerge in water, without the stimulating effect of root diffusates, differs with the species. The larvae of *H. avenae*, contained in the brown cysts, emerge freely in water, if proper temperature conditions are made available and if the 'dormancy' phase is over. The eggs contained in the white cysts (before the cysts turn brown) are capable of immediate hatching, if proper temperature conditions are available. It has been recorded that larvae do emerge from white, yellow or light brown cysts (*H. avenae*) when exposed to 10°C but not if kept at 20-36°C. After the cysts become dark brown, they do not respond to temperature and remain dormant[16,17,19].

In a study, involving six different species of *Heterodera*, emergence of larvae from cysts has been found to be in two phases, active and inactive. Within the active phase, temperature was the influencing factor. The active phase was followed by a period of inactive phase and then again a short active phase. The bulk of the larvae are released during the first active phase[20]. Soil moisture and soil aeration are other important factors in the emergence of larvae from cysts. Maximum emergence from cysts of *H. schachtii* has been recorded at the point of inflexion of the moisture characteristics which, in turn, depends on the soil pore size. It was suggested that the emergence was influenced by moisture and soil aeration[21].

Heterodera avenae, The Cereal Cyst Nematode

The other names, in common usage, for this nematode are : the oat cyst nematode, the cereal root eelworm, 'molya' nematode. The first record of its presence is from Germany[22] in 1874. It is now known to be present in most of the European countries, U.S.A.,

Australia, New Zealand, Canada, Israel, India and Pakistan. The nematode is suspected to be a native of Europe because of recognised variabilities (biotypes) in the populations and also on account of its wide distribution[23]. There are several wheat areas from where there is still no report about its existence. It is possible that the lack of information may be mainly due to absence of qualified nematologists, particularly in the developing countries.

In India, the disease caused by this nematode is referred to as 'molya' and was noticed first in 1954 when causes other than nematodes were attributed to it. This was mainly due to the absence of any qualified nematologist, at that time, in the country. However, there does not appear to be much doubt about the origin of the nematode in Europe because of its earliest record on oats and rye crops, which are believed to have been originated in Europe, with subsequent adaptations on wheat and barley.

Symptoms

The nematode is a root parasite with the second larval stage as the infective one. In the initial stages of infestation, root symptoms are not very specific. Under Indian conditions, elongated main root, with branching of rootlets at extreme ends, is quite often diagnostic. The roots at the extreme ends are bunchy with slight swellings at the points of cyst attachments (Fig. 1). Such symptoms are recognisable within 45 days after sowing. Under European conditions, root division takes place at the invasion points, giving an appearance of knotted roots (Fig. 2). Water stress conditions have been observed to produce similar effects on the roots. Therefore, presence of the cysts on the roots is the only confirmatory indication of nematode infestation[24]. The symptoms on the aerial parts are non-specific. Stunted plant growth, pale-yellowish green leaves with narrow margins are the general features of aerial symptoms. The above ground plant symptoms exhibit patchy growth in field and give an appearance of nitrogen or phosphorus deficiency (Fig. 3).

Host Range

Under field conditions, wheat, barley, oats and rye are recognised as the main hosts for the nematode. Maize is considered as

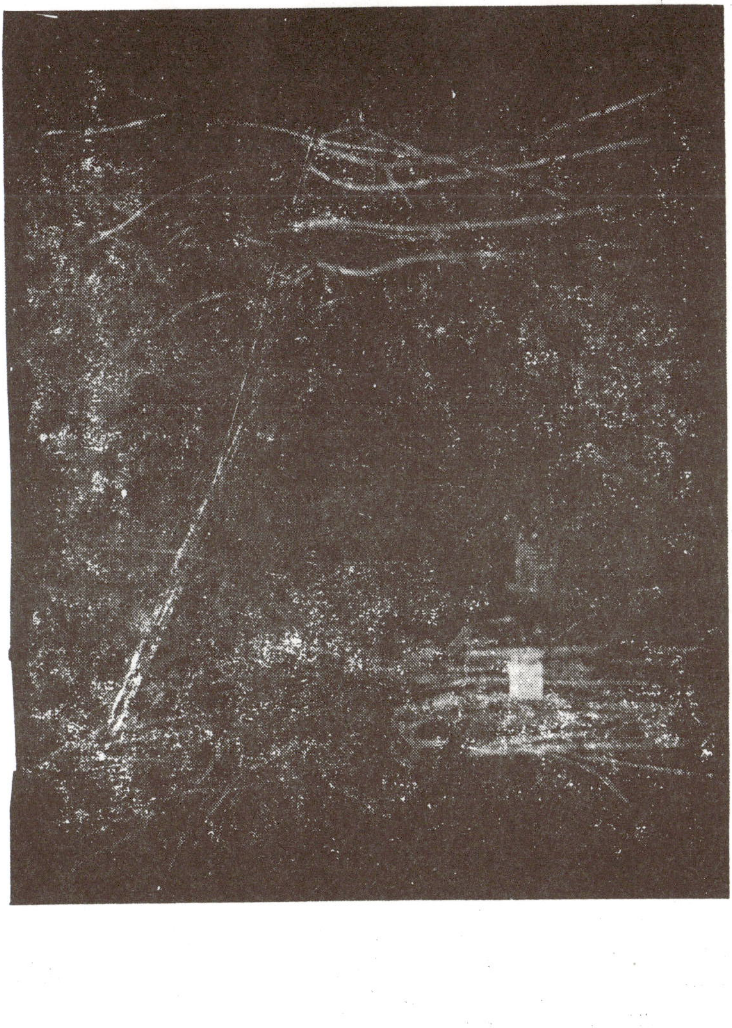

**Fig. 1. Root of wheat seedlings infested with *Heterodera avenae*—elongation of main root with branches at extreme end.
Left—Diseased seedlings ; Righ—Healthy seedlings.**

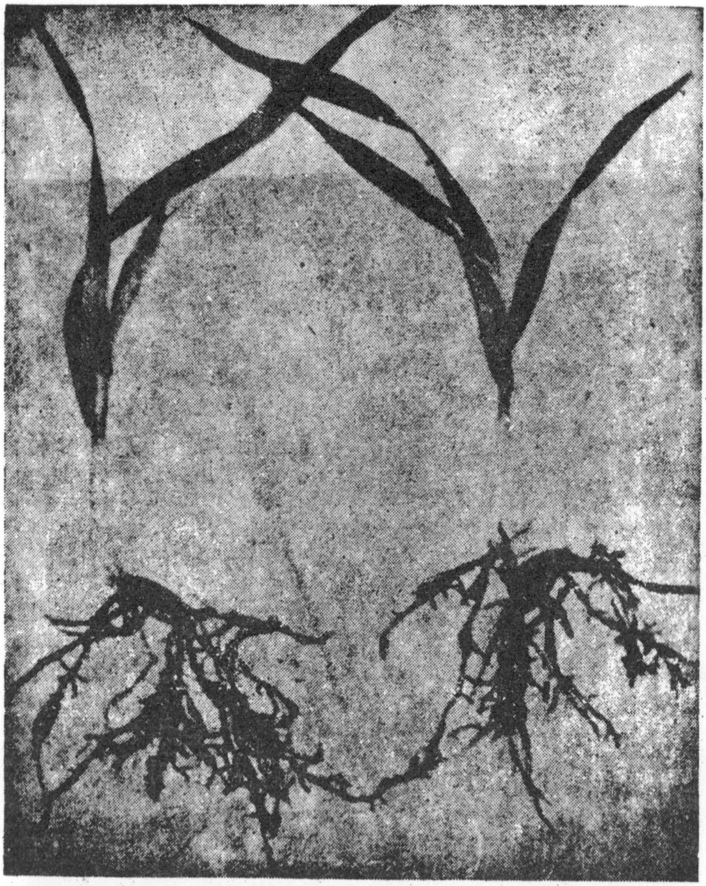

Fig. 2. *H. avenae* on oats.
Left—Healthy seedling;
Right—Roots of diseased plant exhibiting knotting of roots.
(After Hague, 1978).

weak or inefficient host[25,26]. Under pot conditions, *Avena brevis*, *A. byzantina*, *A. sterilis*, *Echinochloa frumantacea*, *Dactylis glomerata*, *Lolium italica*, *L. multiforum*, *L. perenne*, *Phleum pratense*, *Cenchrus ciliaris*, *Phalaris canariensis*, *P. paradoxa*, *Polypogon monospeliensis* (all belonging to Graminae) and *Senebiera pinnatifida* (Cruciferae) have been recorded as hosts[27]. The last named plant is a weak host

Fig. 3. Patchy growth in field infested with *Heterodera avenae*.

and is a common weed in wheat and barley fields of Rajasthan (India) and is the only record outside Graminae. The other grass species recorded as hosts are : *Agropyron repens, Agrostis canina, A. exerata* var. *monolepsis, A. spica-venti, A. stolonifera, Alopecurus pratensis, Ammophila arenaria, A. pratensis, Anisentha maditensis, A. rigida, A. sterilis, Arrhenatherum elatus, Avena abyssinica, A. fatua, A. ludoviciana, A. strigosa, Brachypodium sylvaticum, Bromus arvensis, B. inermis, B. mollis, B. rubens, B. secalinus, B. sterilis, Deschampsia caespitosa, Echinochloa crusgali* var. *frumantacea, Fastuca arundinacea, F. elator, F. longifolia, F. ovina, F. rubra, F. trachyphylla, Hordeum distichum, H. leprinum, H. marinum, H. pallidum, Lolium rigidum, L. temulentum, phalaris arundinacea, P. minor, Poa annua, P. memoralis, P. palustris, P. pratensis, P. trivialis, Secale ancestrale, S. kupryanovii, S. montum, S. segetale, Setaria viridis, Stipa* sp., *Trisetum flavescens* ; other species of *Triticum* like, *T. durum, T. monococcum, T. polonicus, T. Spelta* ; *Sorghum vulgare*[28].

Most of the species are not efficient hosts but are of importance for maintenance of the population in the absence of cereal crops[30].

Biology

The species is amphimictic and follows the life cycle pattern as originally recorded for *H. schachtii*[15]. Irrespective of the geographical region, only one generation of the nematode is known during the crop season. The cysts, which start appearing on the roots a month or so before crop harvest, require a period of 'dormancy' before larvae are activated for emergence. Under Indian conditions, this period is a prolonged one (April to October). It can be shortened though by about two months if proper temperature conditions are made available[30]. Though the optimum temperature for larval emergence is 20°C, emergence can take place at a wide range of 10-25°C. Fluctuating temperatures, however, result in maximum emergence[31,32]. Root exudates (wheat, barley) may also enhance rate of emergence at low temperature (10-15°C) but not so at higher temperatures[32].

Loss Estimates

Nematode population levels present in soil are directly correlated with the damage to the cereal crops. Under conditions, where autumn and spring crops are taken, the same level may not cause equal damage to the different season crops. Under Indian conditions, 20 eggs and larvae/g soil cause significant reductions in wheat and barley growth, with greater damage to barley than wheat[27]. A similar situation prevails in England and Wales also[33].

The economic loss caused by the nematode, under damage threshold population levels, is considered to be quite significant. Most of the loss estimates have been computed indirectly. It has been suggested that for every 10 eggs/g soil, there is a loss of 377 kg/ha in oats, 188 kg/ha in wheat and barley[34,35]. Under Indian conditions 4 and 16 folds increased yields respectively in wheat and barley, after nematicidal treatments have been recorded[36]. In light soils, even at 6 eggs/g soil, the avoidable loss has been calculated as ranging between 45-48 per cent. In terms of monetary loss, the figures are staggering—?,000,000 pounds sterling[37].

Pathotypes

Pioneering work on the identification of the population types was initiated in Denmark[38] with some populations exhibiting resistant reactions on the barley cultivar Drost. Subsequently, 4 pathotypes were found to be present in the European populations[39]. The results of the International Test Assortment, 1972-1973 for *H. avenae*, indicate the possibility of the existence of more pathotypes than hitherto recognised. However, it is quite difficult to demonstrate clearly resistance and susceptibility in terms of number of cysts on the root system. The total number of eggs and larvae per root system appear to be more determinative than the number of cysts. The pathotypes may occur also in mixtures which may complicate the correct identification of the types. This may be one of the possible reasons for different numbers of pathotypes reported by two group of workers in India[40,41].

Control

Though a number of chemicals have been evaluated and found effective also, it has not been possible to apply these under field conditions because of the cost factor. In recent times, some granular chemicals, particularly aldicarb, carbofuran and phenamiphos, have been demonstrated to be quite effective in not only reducing the nematode populations in soil, at the initial level to allow good plant growth, but also have been attributed for increased grain yields. Earlier, DD and DBCP[36] and methyl bromide[42] were demonstrated to be quite effective, but still prohibitive in cost. However, these chemicals, particularly DD and DBCP, have served the useful purpose of assessment of the magnitude of damage.

The other management approaches like cultural (mainly crop rotation and soil amendment), biological and breeding resistant varieties, are promising but still not fully exploited. It has been suggested that wheat and barley should be avoided for atleast two years in heavily infested areas with rotations of non-host crops like beets and alfalfa[43]. Under Indian conditions, a two year rotation with carrots or cumin has proved to be remunerative.

Heterodera schachtii, The Sugarbeet Cyst Nematode

The species has the historical significance as the first cyst-forming nematode identified during 1850 and its association with vast areas of devasted sugarbeet fields in Europe. The nematode was subsequently established as the causal agent of the disease known as 'rubenmudigkeit'. The nematode was named in 1871 as *H. schachtii.* The nematode is widely distributed in all the beet growing areas of the world and is rated as one of the top ten economically important nematode problems in the world.

Symptoms

Like the other nematode problems, the above ground symptoms are in the form of patchy growth of stunted plants, with a tendency to wilt during day hours but recovering during night. There is an excessive development of fibrous root system which carry the cysts.

Host Plants

It is one of the few species which attacks a wide range of host plants in several plant families but mainly Chenopodiaceae and Cruciferae. There are more than 100 plant species, representing 48 genera and 9 families, listed as hosts of the species.

Biology

The cysts are lemon-shaped and possess an egg-sac. There may be 2-5 generations in the crop season depending upon the temperature and vegetative period of the crop growth. Eggs hatch freely in water but can be stimulated further by root exudates from host plants or even by non-host plants like *Hesperis matronalis.* Detailed life cycle has been reported[15] and includes three larval stages outside the egg and adults. The second-stage larvae are the infective stage. 'Giant' and 'dwarf' second-stage larvae are also reported for the species[44].

Control

It is a matter of historical interest that first ever soil fumigation teatment to control a nematode problem was against the sugarbeet cyst nematode in 1881[45]. However, chemical treatments have not given the desired economic returns and therefore other alternative methods like crop rotation with crops like peas, beans, potatoes, etc. or use of quick growing trap crops like mustard or cole seed have been suggested. It is estimated that after a rotation of 3-5 years, one normal crop of beets can be obtained. During the last decade or so, development of granular nematicides and their application either through soil or foliar sprays (phenamiphos and oxamyl) have opened up new possibilities on use of chemicals which may be economically feasible. These chemicals have proved to be quite effective under laboratory conditions.[46]

Heterodera cajani, The Pigeon-Pea Cyst Nematode

The species is quite distinctive from the other species in having a large egg-sac (sometimes almost the double of the cyst size) at the posterior end of the female. Like *H. schachtii*, it has also a wide host range. It can complete several generations, 8-9 in the case of pigeon-pea crop, and thus has a great capacity of quick multiplication and build up of huge populations in a short time.

Host Range

As stated already, the species parasitises a large number of crop plants, mainly belonging to the family Leguminosae and some in Pedaliaceae. Pigeon-pea, green gram, black gram, cowpea and *Sesamum indicum* are recorded as efficient hosts[47].

Biology

Temperature is an important factor governing the time required to compelete the life cycle. At soil temperature of 28.8°C, it takes only 16 days to complete the life cycle as against 45 to 80 days at 10-24°C. Eventhough, it is a bisexual species, it can reproduce parthenogenetically and can complete 4-5 generations in a crop season, even 8-9 in some long duration crops. Males are, however, considered necessary for rapid multiplication.

As in the case of *H. schachtii.* large and dwarf second-stage larvae are recorded for this species too. The optimum temperature for emergence of larvae from cysts is 29°C and the range is 15-37°C.

Distribution

India, Pakistan, Egypt.

Very limited informations are available on the actual losses caused by the nematode. On cowpea, the damaging threshold population level is 0.2 larva/g soil and it also causes reduced nodulation in the crop. Often, nematode infestation is associated with a wilt fungus (*Fusarium udum*) and in such cases as high as 93 per cent wilting of plant has been recorded[48] The traditional pigeon-pea (type host) crop is of about 10 month duration and is often ratooned also. The feeder roots are always available to the larvae and under such conditions, potentials exist for huge population build up. Such populations have been often encountered with plants exhibiting unthrifty growth.

Heterodera zeae, The Maize Cyst Nematode

During the past three years or so, this species has assumed an importance because of possible threat to maize cultivation in U.S.A. The species was first described in 1971[49] from India but is now recorded also from Pakistan, Egypt and U.S.A.

Host Range

The host list is confined to the members of the Graminae family, though some non-graminaceous plants have been recorded also which are of doubtful status. Besides maize (type host), the other listed hosts are : *Hordeum vulgare*, *Triticum aestivum*. *Setaria italica*. *Panicum* species, *Coix lachryma*, *Eleucine coracana*, *Oryza sativa*, *Secale cereale*, *Zea mexicana* and *Vetiveria zizanioides*.

Biology

The nematode thrives well in light soils. Irrespective of the colour of the cysts, emergence of larvae takes place between 20-35°C with optimum at 28°C. The duration of the life cycle is only 17-18 days. In a crop season, there may be 5-6 generations. The species is basically bisexual with limited parthenogenesis[20].

Heterodera trifolii, The Clover Cyst Nematode

It is also one of the species which has a wide host range, mainly confined to the family Leguminosae. It is known to cause reduction of nodules of the attacked plants and also a reduction of about 30 per cent in crude protein content of the plants.

Distribution and Hosts

It is widely distributed in Europe, Canada and U.S.A. and in restricted areas of India and New Zealand. *Trifolium repens*, white dutch clover, is a good host but 'ladino clover' (*T. repens*) and 'red clover' (*T. pratense*) are poor hosts. The other good hosts are: *Pisum sativum, Chenopodium glaucum, Vicia villosa, V. narbornesis, Rumex crispus, R, nemarosa, Isatis tinctoria, Stellaria media, Sesbania macrocarpa, Dianthus caryophyllus* and carnations.

The nematode takes 26 days to complete the life cycle at a soil temperature of 15°C.

The list of the *Heterodera* species has grown enormously during the past few years and it is not possible to mention all of them in some detail and therefore, the above descriptions have been limited to some of the important ones which may be considered as representatives of the group. A list of the other described species is given below, some of these are also associated with economically important crops

List of other species of *Heterodera* : *H. acnidae, H. amygdali, H. arenaria, H. bifenestra, H. canadensis, H. cardiolata, H. carotae, H. cruciferae, H. cyperi, H. daverti, H. delvii, H. elachista, H. fici, H. filipjevi, H. galeopsidis, H. gambiensis, H. glycines, H. goettingiana, H. graduni, H graminis, H. graminophila, H. hordecalis H. humuli, H. iri, H. lespedezae, H. leuceilyma, H. limoni, H. longicola, H. mani, H. medicaginis, H. mediterranea, H. menthae, H. methwoldensis, H. mothi, H. oryzae, H. oryzicola, H. oxiana, H. paratrifolii, H polygoni, H. raskii, H. rosii, H. rumicis, H. sacchari, H. salixophila, H. scleranthi, H. sonchophila, H. sorghi, H. tadshikistanica, H. urticae, H. ustinovi, H. uzbekistanica.*

Globodera rostochiensis and *G. pallida*, Golden Nematode of Potato

Amongst the cyst-forming nematodes, these two species of nematodes are considered as one of the most devastating nematode problems of the world. Its association with potato was recognised almost at the same time as the beet cyst nematode without any attempts for its identification. It was found to be present in Europe over large areas and at places caused serious damage to the potato crop. It was only in 1923, Wollenweber[4], while examining cysts obtained from a place, Rostock, in Germany, established this as a species different from *H. schachtii* and named it as *H. rostochiensis*. Because of its economic importance, it has received considerable attentions from the European as well as American nematologists. One of the most important research thrusts has been towards developing resistant lines. During the course of such investigations, it was observed that there are certain populations of the nematode which have the capacity of breaking down the resistant lines developed through breeding processes. It was ultimately resolved that the populations comprised of two species, with one of them having golden yellow maturing females (*rostochiensis* group) and the other having white or cream colour (the specific name *pallida* was given to it). Later, mainly because of the round shape of the cysts, the sub-genus *Globodera* was given the full generic level and accordingly the two species have been named as *Globodera rostochiensis* and *G. pallida*. The earlier literature on the potato cyst nematode, therefore, has to be viewed with some caution.

The first record of the presence of the potato cyst nematode on the Long Islands (U.S.A.) is considered as one of the important land-marks in the development of nematology in the world. The discovery initiated several administrative actions requiring trained nematologists, which, in turn, paved the way for organised teaching of the subject in the universities. The name 'golden nematode of potato' was proposed in 1944 by Chitwood[50].

Distribution

It has been practically recorded from most of the countries of the world, with vast areas of potato fields affected by its infestation in Europe. In U.S.A., intensive efforts are being made to keep it confined to the Long Islands. The same is true of the limited infestation pocket of Nilgiri Hills in India. It has been reported also from Peru (South America) which is considered to be home of potatoes. It is suspected that the original home of the nematode is also Peru.

Biology

The life cycle processes are similar to the other cyst-forming species, with one life cycle being completed in 5-7 weeks. This is mainly a temperate area problem and thrives well under cool conditions. The eggs and larvae present inside the cyst body require a stimulation from host root diffusate for hatching and emergence. In the absence of potato crop, there is very little emergence of the larvae. The natural decline in population, under such circumstances, is considered to be between 30-50 per cent per annum[51]. However, those which do not emerge, can remain in viable condition inside the cysts even for 7-8 years. Even when the potato crop is available, all the larvae do not emerge out of the cysts and may remain, without losing viability, in the cysts for some years.

Symptoms

As with other nematode infestations, patchy growth of stunted plants is the main indication (Fig. 4). Under severe infestation conditions, the plants tend to wilt during the day hours but recover at night. The roots do not exhibit full development and may become at brown. There may be increased branching of the root system with cysts attached to them (Fig. 5). Initially, the young seedlings, while drawing nutrition from the tubers, do not show any signs of injury, but after the seed tuber has dried up and the plant starts drawing nutrients through the roots, the symytoms become pronounced. Severely infested plants contain less K and P but more Ca and Na. Nitrogen status is little affected[62,63].

Hosts

Potato is the main host but tomato, brinjal and snapdragon are also susceptible.

Control

Elimination of potato cyst nematodes from field soils is difficult because of its persistence and large multiplication rates at low infestation levels. Therefore, lots of efforts are being put for developing management schedules, incorporating one or more methods of control, which would help to bring down the nematode population

Fig. 4. Patchy growth of potato field infested with *Globodera rostochiensis*. (After Hague. 1978).

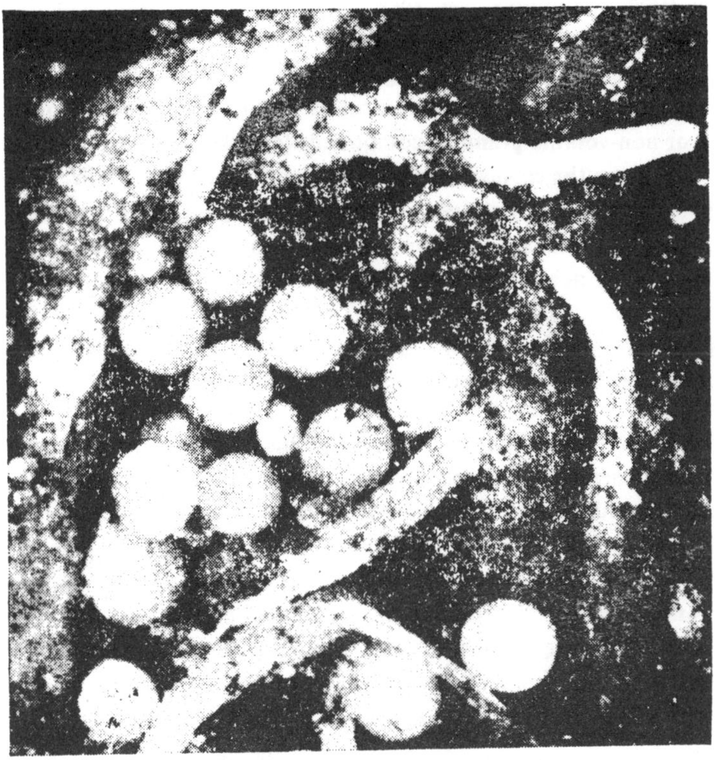

Fig. 5. Cysts attached to roots.
(After Hague, 1978).

below the damage level and still have the cost-benefit ratio in favour of the farmer. Apart from strict regulations for enforcement of quarantine, sanitation (a thorough washing of the seed potatoes to remove adhering cysts ; treatment of seed potatoes with NaOCl or CaOCl), crop rotation with atleast 4 years intervals between two potato crops, growing of resistant varieties (CP 1422, 1781, 1806) are quite effective measures, singly or in combinations of each other or chemicals. It has been demonstrated that incorporation of resistant variety alone in a 4 year crop rotation programme can result in 67-78 per cent yield increase[54]. The chemicals, by themselves, though effective are quite expensive and therefore, have to be used judiciously for economising on the cost by modifying the application technology. A very good control of the nematode problem has been obtained when aldicarb or oxamyl are incorporated on the ridge soil before planting susceptible potatoes rather than mixing in field soil. Chemical control by fumigants (telone, dazomet, etc.) or non-volatile granular compounds (aldicarb, oxamyl, etc.) is possible where the economics of the crop makes it worthwhile. The present day strategy, however, is to use combination of crop rotation, chemicals and resistant varieties. Early potatoes may also escape damage by the nematode as the crop may be harvested before the nematode completes its life cycle. Under European conditions, a single generation from April to July is normal but there may be a second or partial second generation in warmer soils.

One of the major sources of spread of the cyst populations to new uninfested areas is transport of the infested soil through farm machinery, animal or human feet or strong winds. It has been possible to recover 32,000 cysts from tractor wheels operating in an infested area[60]. These cysts can survive for quite some time and therefore, once introduced in an area, they will persist and multiply, if suitable hosts are available. To control the nematode, therefore, legislative action prohibiting movement of seed potatoes or non-use of contaminated farm machinery, once used in the infested area, must be taken to safeguard the interests of the nation.

The other recognised species under the genus *Globodera* are: *G. artemisiae*: *G. leptonepia*, *G. mali*, *G. millefolii*, *G. mirebilis*, *G. pseudorostochiensis*, *G. solanacearum*, *G. tabacum*, *G. virginiae* and *G. zelandica*. Except perhaps *G. tabacum*, the economic importance of other species is uncertain. *G. tabacum* is quite close to *G. rostochiensis* but apart from some morphological differences, it does not mature on potato but does so on other hosts of *G. rostochiensis*. The potato cyst nematode also does not go to tobacco.

OTHER GENERA OF CYST NEMATODES

Cactodera, *Ephippiodera*, *Punctodera* and *Sarisodera* are the other genera which are cyst-forming. However, their importance to economically important crop plants is not known.

REFERENCES

1. Schacht, H., Ueber einige Feinde und Krankheitan der Zuckerrube. *Zeitschr. Ver. Rubenzucher. Ind. Zoolver. 9*, 390, 1859.

2. Schmidt, A., Ueber den Ruben-Nematoden (*Heterodera schachtii*). *Zeitscher Ver. Rubenzucker-Ind. Zoolver. 22*, 67-75, 1871.

3. Liebscher, G., Beobachtungen uber das Auftreten eines Nematoden an Erbsen. *J. Landwirts. 40*, 357-368, 1892.

4. Wollenweber, H., Krankheiten und Beschadigungen der Kartoffel. *Arb. Forsch. Inst. Kartoff. 7*, 52, 1923.

5. Franklin, M.T., On the specific status of the so called biological strains of *Heterodera schachtii. J. Helminth. 18*, 193-208, 1940.

6. Golden, A.M., Classification of the genus and higher categories of the Order Tylenchida (Nematoda). In : Plant Parasitic Nematodes, Eds. Zuckerman, B.M., Mai, W.F. and Rohde, R.A., *Acad. Press*, New York, 191-230.

7. Wouts, W.M. and Sher, S.A., The genera of the sub-family Heteroderinae (Nematoda : Tylenchoidea) with a description of two new genera. *J. Nematol. 3*, 129-144, 1971.

8. Krall, E.L. and Krall, H.A., Systematic changes of phytonematodes of the family Heteroderidae based on trophic specialisation of these parasites and their evolutionary connection with agricultural plants. *Phytohelm. Proc. Akad. Nauk., c.c.c. P. Moskow*, 39-56, 1978.

9. Mulvey, R.H. and Ebsary, B.A., *Dolichodera fluvialis* n. gen., n. sp. (Nematoda : Heteroderidae) from Quebec, Canada. *Canad. J. Zool., 38*, 1697-1702, 1980.

10. Behrens, E., *Globodera* Skarbilovich, 1959, eine selbstandige Gattung in der unterfamilie Heteroderinae Skarbilovich, 1947 (Nematoda : Heteroderidae) 1. Vortragstagung Zu Aktuellen Problem der Phytonematologie am 1975 in Rostock. *Manuscript druck der Vartrage. Rostock*, 12-26, 1975.

11. Mulvey, R.H. and Stone, A.R., Description of *Punctodera matadorensis* n. gen., sp. (Nematoda: Heteroderidae) from Saskatchewan with lists of species and generic diagnoses of *Globodera* (n. rank), *Heterodera*, and *Sarisodera*. *Canad. J. Zool.*, *54*, 772-785, 1976.

12. Oostenbrink, M., and den Ouden, H., De structuur van de kegeltop als taxonomisch kenmerk bij *Heterodera* soorten met citroenvormige cysten. *Tijdschr. PlZiekt.*, *60*, 146-152, 1954.

13. Cooper, B.A., A preliminary key to the British species of *Heterodera* for use in soil examination. *Soil Zool. Proc. Univ. Nottingham*, 269-280, 1955.

14. Swarup, G., Sethi, C.L, Kaushal, K.K. and Siyanand, Distribution of *Heterodera avenae*, the causal organism of 'molya' disease of wheat and barley in India. *Curr. Sci.*, *51*, 896-897, 1982.

15. Raski, D.J., The life history and morphology of the sugarbeet nematode, *Heterodera schachtii* Schmidt. *Phytopathology*, *40*, 135-152, 1950.

16. Banyer, R.J. and Fisher, J.M., Seasonal variation in the hatching of eggs of *Heterodera avenae*. *Nematologica*, *12*, 225-226, 225-236, 1971.

17. Banyer, R.J. and Fisher, J.M., Effect of h t and environment on some aspects of the biology of *Heterodera avenae*. *Nematolodica*, *22*, 298-305, 1976.

18. Okada, T., Studies on the hatching factors of the soybean cyst. nematode, *Heterodera glycines* ichinohe. *Japanese J. Nematol.*, *6* (Special issue), 50 pp., 1977.

19. Rajan, Evaluation of cyst extraction techniques and their effect on biology of *Heterodera cajani*, *H. zeae* and *H. avenae*. Ph.D. Thesis, IARI, New Delhi, 93 pp., 1984.

20. Sharma, S.B. and Swarup, G., Cyst forming nematodes of India. Cosmo Publ., New Delhi, 152 pp., 1984.

21. Wallace, H.R., Factors influencing the emergence of larvae from cysts of beet eelworm, *Heterodera schachtii* Schmidt. *J. Helminth.*, *29*, 3-16, 1955.

22. Kuhn, J., Ub'r has Vorkommen von Ruben-Nematoden an den Wurzein der Halmfruchte. *Landwirts. Jahrb.*, 3, 47-50, 1874.

23. Meagher, J.W., World dissemination of the cereal cyst nematode (*Heterodera avenae*) and its potential as a pathogen of wheat. *J. Nematol.*, *9*, 9-15, 1977.

24. Kort, J., Nematode disease of cereals in temperate climates. In : Economic Nematology, Ed. J.m. Webster, Acad. Press, 97-108 pp.

25. Swarup, G., Prasad, S.K. and Raski, D.J., Some *Heterodera* species from India. *Plant Dis. Reptr.*, *48*, 235, 1964.

26. Johnson, P.W. and Fushtey, S.G., The biology of the oat cyst nematode, *Heterodera avenae*, in Canada. II. Nematode development and related anatomical changes in roots of oats and corn. *Nematologica*, *12*, 630-636, 1966.

27. Gill, J.S. and Swarup, G., On the host range of the cereal cyst nematode, *Heterodera avenae*, the causal organism of 'molya' disease of wheat and barley in Rajasthan, India. *Indian J. Nematol.*, *1*, 63-67, 1971.

28. Holdeman, Q.L. and Watson, T.R., The oat cyst nematode, *Hererodera avenue*, a root parasite of cereal crops and other grasses. *Rept. State of California*, U.S.A., Dept. Food and Agric., 82 pp., 1977.

29. Stone L.E.W., Field experiments on cereal root eelworm. *Ann. appl. Biol.*, *48*, 681-686, 1960.

30. Gill, J.S., Studies on Heterodera avenae, the causal organism of 'molya' disease of wheat and barley. Ph.D. Thesis, Panjab Univ., India, 1967.

31. Swarup, G. and Gill, J.S., Factors influencing emergence of larvae from cysts of *Heterodera avenae*. *Indian J. Exptl. Biol.*, *10*, 219-223, 1972.

32. Williams, T.D. and Beane, J., Effect of temperature and root exudates on *Heterodera avenae* hatch. *Rept. Rothamsted Expt. Sta.* for 1971, *Part I*, 171-72, 1972.

33. Stone, L.E.W., Cereal cyst nematode in spring barley damage assessment experiments, 1963-1964. *Plant Pathology, 17*, 145-150, 1968.

34. Duggan, J.J., The effect of cereal root eelworm (*Heterodera major*) on its host. *Irish J. Agric. Res., 1*, 7-16, 1961.

35. Dixon, G.M., The effect of cereal cyst eelworm on spring sown cereals. *Plant Pathology, 18*, 109-112, 1969.

36. Swarup, G., Mathur, R.L., Seshadri, A.R., Raski, D.J. and Mathur, B.N., Response of wheat and barley to soil fumigation by DD and DBCP against 'molya' disease caused by *Heterodera avenae. Indian J. Nematol., 6*, 150-155, 1976.

37. Wallace, H.R., The ecology and control of the cereal root nematode. *J. Aust. Inst. Agrc. Sci., 31*, 178-186, 1965.

38. Andersen, S., Resistance of barley to various populations of the cereal root eelworm (*Heterodera major*). *Nematologica, 4*, 91-98, 1959.

39. Kort, J., Dantuma, G., and Van Essen, A., On biotypes of the cereal root eelworm (*Heterodera avenae*) and resistance in oats barley. *Neth. J. Plant Path., 70*, 9-17, 1964.

40. Mathur, B.N., Arya, H.C., Mathur, R.L., and Handa, D.K., The occurrence or the biotypes of the cereal cyst nematode (*Heterodera avenae*) in the light soils of Rajastan and Haryana, India. *Nematologico, 20*, 19-26, 1974.

41. Swarup, G., Sethi, C.L., Seshadri, A.R. and Kaushal, K.K., On the biotypes of *Heterodera, avenae*, the causal organism of 'molya' disease of wheat and barley in India. *Indian J. Nematol., 9*, 164-168, 1979.

42. Kyrou, N.C., Chemical control of cereal root eelworm,

Heterodera avenae, in field soil. *Plant Dis. Reptr.*, *54*, 134-136, 1970.

43. Kort, J., Some observations on soil population of cereal root eelworm, *Heterodera avenae* (=*H. major*) under the influence of various crops in sandy soils. *Tijdschr. Pl. Ziekten.*, *65*, 1-4, 1959.

44. Johnson, R.N. and Viglierchio, D.R., Incidence of aberrancy in *Heterodera schachtii. Nematologica*, *16*, 33-38, 1970.

45. Kuhn, J., Die Ergebnisse der Versuche zur Ermittelung des Ursache der Rubenmudigkeit und zur Erforschung der Natur der Nematoden. *Ber. Physiol. Lab. Univ. Halle.*, *3*, 1-153, 1881.

46. Griffin, G.D., Control of *Heterodera schachtii* with foliar application of nematicides. *J. Nematol.*, *7*, 347-351, 1975.

47. Koshy, P.K,, and Swarup, G., Susceptibility of plants to the pigeon pea cyst nematode, *Heterodera cajani. Indian J. Nematol.* *2*, 1-6, 1972.

48. Hasan, A., A synergism between *Heterodero cajani* and *Fusarium udum* attacking *Cajanus cajan. Nematologica Medit.*, *12*, 159-192, 1984.

49. Koshy, P.K., Swarup, G. and Sethi, C.L., *Heterodera zeae* n. sp. (nematoda : Heteroderidae), a cyst forming nematode on *Zea mays. Nematologica*, *16*, 511-516, 1971.

50. Chitwood, B.G., The golden nematode of potatoes. *U.S. Dept. Agr. Circ.*, 875.

51. Huijsman, C.A., The influence of resistant potato varieties on the soil population of *Heterodera rostochiensis. Nematologica*, *6*, 177-180, 1961.

52. Trudgill, D.L., Evans, K., and Parrot, D.M., Effects of potato cyst nematodes on potato plants. I. Effects in a trial with irrigation and fumigation on the growth and nitrogen and potassium contents of a resistant and a susceptible variety. *Nematologica*, *21*, 169-182, 1975.

53. Trudgill, D.L., Evans, K., and Parrot, D.M., Effects of potato cyst nematodes on potato plants. II. Effects on haulm size, concentration of nutrients in haulm tissue and tuber yield of a nematode resistant and a nematode susceptible potato variety. *Nematologica, 21*, 183-191, 1975.

54. Zawislak, K., Sowa, S., and Gronowicz, Z., Wplyz matwiko-odpornych odmiar Ziemniaka ha zenikanie w glabie populacji matwika ziemniaczanego (*Globodera rostochiensis*). *Zeszyty Naukowe Akademi Rolniczo-Technicznoj w olsztynie*, Rolnictwo, No. 31, 49-58, 1981.

11

Genus Rotylenchulus Linford and Oliveira, 1940

The nematode was first discovered on cowpea in 1931 in the Island of Oahu, Hawaii, U.S.A.[1], but was not given any name. About five years later, it was recorded again on the roots of cowpea, raised in a pineapple field of Oahu county. Subsequent studies led to its description under *Rotylenchulus reniformis*. The generic name was proposed to indicate certain morphological similarities of the nematode with the genus *Rotylenchus* as well as close resemblance of its life cycle with the genus *Tylenchulus*. The specific name was derived to reflect on typical kidney shape of mature females of the nematode. Since the first description, nine more species have been added under this genus. A major revision of the genus was carried out in 1968[2] giving out detailed descriptions of topotye and several other populations, collected from various parts of the world.

Morphology

The species are characterized by sexual dimorphism. The mature females are swollen to kidney shape and males remain vermiform, with poorly developed stylet and oesophagus. Immature females, males and larvae have four incisures in the lateral field. Posteriorly, in males, these incisures extend to varying distances, from slightly anterior to the cloacal opening to about midway on the tail. The excretory pore is located posterior to the median bulb in all species. The hemizonid is two to three annules long and immediately anterior to the excretory pore.

The lip region is continuous, not set off and varies in shape and annulation. Most species have either low and rounded lip region (*R. parvus*) or higher and conoid in shape (*R. macrodoratus*). Some species appear to be intermediate between these two categories (*R. reniformis*).

Stylet length in immature females ranges from 10-26 μm. The stylet knobs slope backwards except in *R. macrodoratus* where knobs are anchor shaped. The male stylet is smaller and less developed than the corresponding female stylet and knobs are much reduced. The location of the dorsal oesophageal gland orifice varies from 0.56-1.90 times the stylet length, posterior to the stylet knobs, and is diagnostic for different species.

The oesophagus of mature females is long with a well developed median bulb and valve. The posterior region of the oesophagus is usually hemispherical in shape but there is occasional inter-population variations in this character. The glandular region of the oesophagus overlaps the intestine laterally and ventrally.

The females have didelphic, amphidelphic ovaries. The ovaries of immature females are very small and undeveloped. One of the useful characters in distinguishing species is the post-median position of the vulva. The value of V falls usually between 58-66 except for *R. reniformis* and *R. onamictus* where it is more posteriorly located. The male reproductive system is monorchic and outstretched. Gubernaculum and spicules are well developed. Caudal alae are much reduced, adanal and annulated.

Generally, the tail shape of the larvae is rounded whereas it is more pointed in the immature females. There is considerable intraspecific variation in tail shape of immature females.

The distal portion of the tail of mature females, in some species, is set off into a finger-like appendix. In other species, the the distal portion of the tail is in the shape of a short spike. Still in other species, the tail is fluently rounded with no indication of an appendix or projection.

The best studied species is *R. reniformis* because of its widespread occurrence and potentiality to cause damage to many cultivated crops of economic importance. The nematode is usually confined to tropical and subtropical areas of the world and is reported to occur in more than 30 countries including Brazil, Sri Lanka, Cuba, India, Japan, Indonesia, Iran, Jamaica, Liberia, Ghana, Mexico, Nigeria, Bangladesh, Philippines, U.S.A., Venezuela, Western Panama, Angola, Somalia, British Honduras, Fiji, Malawi, Spain, Greece, Malta, Israel, Egypt, Kenya, Lebanon, China, Pakistan, Dominican Republic, Puerto Rico[2]. In India, the nematode has been reported from all the states including Punjab. Haryana, Rajasthan, Uttar Pradesh, Bihar, Tamil Nadu and Karnataka[3]. In U.S.A., the distribution is confined to Hawaii, Alabama, Texas, Louisiana, South Carolina and California[1,4–8]. There are some evidences that the nematode can survive in cold climates as well. The possibility exists, therefore, that its distribution may be beyond tropical and sub-tropical areas of both the hemispheres. However, because the thermal limits for reproduction are between 15 and 36°C, there is little likelihood that it may cause any serious concern to crops in temperate regions.

Host Range

It is polyphagous and is known to feed and multiply on a wide range of horticultural, vegetable and plantation crops. Recent studies indicate that more than 115 plant species, mostly belonging to Leguminosae, Curubitaceae, Malvaceae and Solanaceae serve as hosts for this nematode[8,10,11]. Some of the crops on which noticeable injuries have been recorded are : cotton, papaya, pineapple, cowpea, castor bean, banana, tea, sweet potato, coffee, pigeonpea, tomato, tobacco, okra, eggplant, beans, soybean and cantaloupe.

Biology

Amongst the tylenchids, the life cycle of the reniform nematode is rather unusual in having the infective and feeding stages confined to immature females only. Another unique feature is the transformation of second-stage larvae into immature females and males, through three superimposed moults without feeding. The duration of different developmental stages and fecundity are influenced by environmental factors like temperature and host[1,12-14].

In a natural population, the ratio of male to immature female is 1 : 1 and the nematode reproduces amphimitically. Parthenogenesis is considered to be a rare occurrence[15]. However, it has been reported in three populations from Japan and India[14,16]. The anterior portion of the immature females penetrate the host roots inter or intracellularly. They orient themselves almost perpendicularly to the central axis of the infected root and start feeding semi-endoparasitically. The posterior part of the body, which remains outside the root, begins to swell within 3-5 days of post-infectional period and in course of time assumes the typical kidney shape. The matrix secreting glands, consisting of six unicellular pear shaped cells, on either side of the viginal canal, secrete gelatinous matrix, which strongly attracts the males. Many males are often observed to be coiled within the gelatinous matrix. About 8-10 days after penetration of roots, the females start depositing single celled eggs. The total number of deposited eggs varies from 39-120, depending upon hosts and other environmental factors. The life cycle duration is reported to be 25 days, from egg to egg, on cowpea[1] ; 24-29 days on okra[14] ; 17-22 days on cotton[18] ; 20 25 days on egg plant[17] and 14 days on tomato[18].

Soil moisture and temperature have profound influence on infectivity, penetration and the biology of the nematode. A combination of 25-30 per cent soil moisture and 30°C is considered as optimum for reproduction of the nematode[19]. The optimum moisture level for survival is reported to be 28.6 per cent[20]. However, it can survive, to some extent, in air-dried soil (3.3 per cent moisture) for 7 months, if stored at 20-25°C[21]. Under Hawaii (U.S.A.) conditions, it has been recorded to survive better at low moisture levels than high which suggested that moist fallow might be an effective method in reducing the populations[22].

The optimum temperature for penetration and development of the nematode appears to be different. In a study with cantaloupe, it has been seen that although the nematode penetrated the roots in equal numbers at 21 and 27°C, within 24 hours after inoculations, further development was better at high temperature (27°C) than at the lower one (21°C)[23]. The influence of temperature on life cycle was demonstrated further in a study using resistant and susceptible soybean cultivars. In both the cultivar types, maximum numbers of population was encountered at 29.6°C than at 15.0, 21.0 or 36.0°C. In susceptible cultivars, the nematode completed its life cycle in 19 days at 29.6°C while the females developed without producing egg masses at 15.0 or 36.0°C[24]. Soil pH is also an important factor affecting reproduction of the nematode. Generally, it thrives best in slightly acidic soil, with optimum between 4.8-5.2[25]. However, even under saline conditions, the nematode has been observed to reduce plant (cotton) growth. Perhaps, limited development of nematode, is due to poor root growth to support enough population, under saline conditions, rather than any adverse effect on the nematode itself[26].

The existence of physiological populations of the nematode was suspected during the early sixties[27,28] but a positive demonstration was made only in 1971 on the basis of parasite's reproductive capacity on differential hosts[29]. Populations which reproduced on cowpea, castor and cotton have been designated as Race A while one population which reproduced only on cowpea but not on cotton and castor has been named as Race B. The two races interbreed with each other and the hybrids are more polyphagous than either of the parents. Consecutive inoculations of the original population onto a poor host plant, up to 12 generations, have resulted in an evolution of an aggressive population, which reproduced very well on original poor host. On the basis of these observations, it has been suggested that the races probably evolved through host specialization and subsequent introgressive hybridization of host specialized populations[15].

It is likely that *R. reniformis* causes economic damage to a wide range of crops in many tropical and subtropical areas of the world. Nonetheless, substantial data are yet to be generated on the quantitative as well as qualitative losses of crop yields. As in many other cases, the economic damage of the nematode has been computed on the basis of fumigation trials. In cantaloupes, fumigation resulted in 79-105 per cent higher yields than the unfumigated plots. It was calculated also that there was a gross increase in monetary benefits (1965 dollars/ha) from increased yields in comparison to fumigation cost of only 20 dollars/ha[30]. Likewise fumigated plants exhibited significant increase in sucrose compared to non-fumigated crops.

Perhaps, maximum research attention on this nematode has been given in respect to cotton. Under field conditions, the nematode has been observed to cause an appreciable reduction in yield in all varieties of cotton, with fumigated plots outyielding the checks by 22.4-79.6 per cent[31]. The nematode also caused a delay in maturity as well as reduction in boll size and lint percentage. It is reported also to cause an increase in wilt development in wilt susceptible cotton varieties.

Soybean is another important crop which is very much affected by its infestation[32-34]. The yields may be reduced by as much as 45 per cent besides also reducing the phosphorus content of seed. The nematode is also reported to cause a significant reduction in yield of sweet potato by 60 bushels per acre. Application of chemicals, under field conditions, have resulted in significant increases in yield of US 1 sweet potato grade by 605, 615, 710 and 905 kg/ha during the trial years.

There ars only few records to indicate the qualitative loss in crops due to *R. reniformis*. The loss in oil content of castorbean has been reported to be 6.39, 11.07 and 15.63 per cent from plants infected with the nematode at an initial inoculum levels of 200, 2000 and 20,000/10 1 of soil respectively[35]. Similarly, the nematode has been recorded to cause significant reduction in marketable size of some potato cultivars[36].

Symptoms

The gross symptomatology of reniform infected plant species has been studied for few crops only. When castorbeans are heavily infected with the nematode, drying of leaf margins are noticed within one month of post-infectional period[35]. The drying of leaves spreads gradually to lamin followed by premature shedding of leaves (Fig. 1)

The axillary buds in main stem develop into tiny leaves and later shed off, leaving the plants only with stem. The roots show moderate discolouration with the root-lets as dark brown and sometimes necrotic. The seeds from nematode-infected plants are malformed and discoloured. In the tender roots of tomato and pigeonpea, traces of necrotic tissued are regularly observed around the female heads (Fig. 2). A severe curling of young roots of many crops are observed in the case of heavy infection. On cotton, few small and less dark green leaves are noticeable in nematode stressed plants. Infected cotton plants remain stunted with less vigour. Severe infection often results in pruning of roots[81].

Hypertrophy, hyperplasia, thickening of cell walls, dense granular cytoplasm, enlargement of nucleus, formation of giant cells, syncytia, feeding peg, feeding tube, extensive mechanical damage to cortical tissues are usually observed, in variety of crops infected with *R. reniformis* (Fig. 3).

In cotton, immature females initiate infection by extending anterior portion of body through epidermis, cortex, endodermis and pericycle of young host root before coming to rest and feeding at phloem[13]. The damage to phloem cell, as a result of feeding, extends

Fig. 1. Drying of castor leaves infected with *Rotylenchulus reniformis*. Left—Healthy plant ; Right—Infected plant.

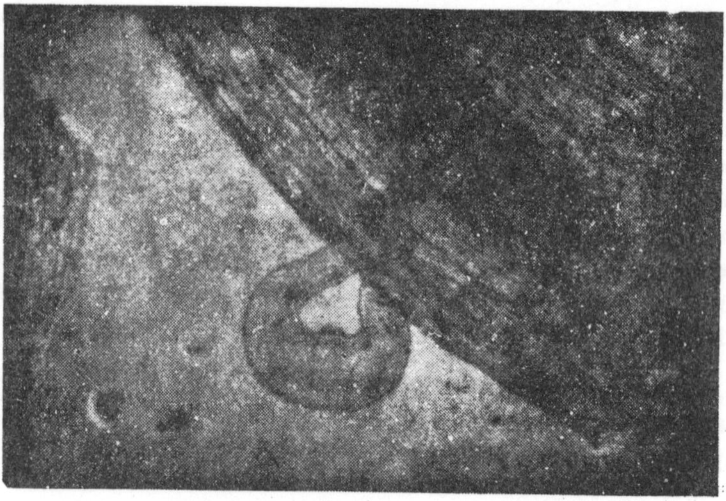

Fig. 2. Necrotic tissues around nematode head.

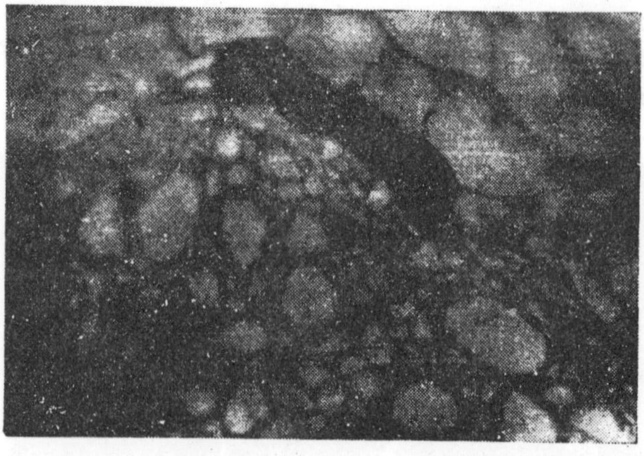

Fig. 3 Extensive damage to cortical tissues as a result of *Rotylenchulus reniformis* infection.

to several cells along the axis of roots. Heavy infection results in damage to epidermal and parenchymal tissues. Severe injury may lead to necrosis of phloem and parenchyma and subsequent dwarfing of cotton plants. *R. reniformis* infected-tomato and papaya suffer similar damage[37]. Recently, it has been demonstrated that the immature females penetrate the cortex of cotton roots intracellularly until its labial disc makes contact with an endodermal cell[53]. Usually penetration is perpendicular to root axis. Approximately half of the conical portion of stylet is inserted into prosyncyte (PCS) or initial syncytial cell (ISC) which is usually an endodermal cell, to initiate feeding. About 11 days after infection, the cell wall material are synthesized and deposited by ISC in response to stylet penetration and nematode feeding, to form feeding peg which covers the stylet tip. The feeding peg is made mostly of plant derived material and benefits the nematode development by preserving the integrity of syncytium, through prevention of excessive leakage of cytoplasm or cytoplasmic content or entry of other foreign substances in syncytium. The lumen of the stylet is continuous with that of the feeding tube at the stylet aperture. Both feeding tube as well as the peg are associated with feeding of reniformis.

The penetration and development of the nematode cause extensive mechanical change to the cell wall of soybean roots[38]. The nematode comes to rest with anterior end, inside pericycle cell, which is transformed into a giant cell, to serve as feeding site. As many as 100-200 adjoining giant cells may be associated with the feeding of one female. The females that feed on pericycle tissue, adjacent to xylem vessels of protoxylem poles, but not protophloem poles, are able to complete life cycle with production of egg mass and giant cell. Ultrastructural studies on feeding mechanism of the nematode on cotton roots indicate that the females penetrate the root cortex intracellularly and establish PSC or ISC to initiate feeding in an endodermal cell. Emanating from the stylet aperture is the tubular secretion to form feeding tube as observed in soybean and cowpea infected with the nematode. Also, another structure associated with feeding is feeding peg, in the cases where the tip of stylet is inserted into ISC. At the early stage of feeding peg development, membranous paramural bodies and fragmented cell wall like material are seen on cell wall side of plasmalemma. Feeding peg, a structure of plant origin appears to be biochemically polysaccharides and lignin[39–42].

The feeding-site in cantaloupes also appears to be an endodermal cell[43]. Feeding by nematode stimulates the pericycle tissues, on either side of the endodermis and cause cellular hypertrophy with enlargement of nucleoli and granular thickening of the cytoplasm. In the feeding cell, a small digit like indentation surrounds the stylet. In the older roots where endodermis has collapsed, the nematode feeds on pericycle. In sunflower, the feeding site could be of pericycle origin[44]. Cortical layer adjacent to the endodermis and the endodermis itself can serve as the feeding site. The affected cells are uninucleate in majority of the cases and PSC are endodermis. In contrast to soybean, normal female development occurs even when they feed on cortical or pericycle cells. The disruption of endodermis, needed by the nematode to reach pericycle, may impair the osmotic stability of vascular cylinder.

In susceptible soybean cultivars, two basic phases of development have been observed :

(1) an anabolic phase characterized by proliferation of organelles accompanied by secondary wall deposit to provide nutrition to developing females ; and

(2) an initial phase marked by partial cell wall lysis and separation of cell.

The resistant cultivars lack anabolic phase and exhibit hypersensitive reaction.

The association of reniformis with other micro-organisms like fungal pathogen, often results into disease complexes, where damage expression or crops may be additive and severe (Table 1). Such a situation had been visualised in case of cotton wilt in the early forties[60,61]. It was subsequently recognised that the nematode is an important component of a disease complex in cotton where it enhances significantly *Fusarium* wilt incidence in wilt susceptible cotton cultivars but not in wilt resistant cultivars[45-46,62-63]. Similarly, infection of soybean with reniformis nematode has been found to increase the replications of cowpea chlorotic mottle virus.

Control

Chemical

The discovery of D-D, as a nematicide, led almost the entire pine apple industry of Hawaii to adopt a control programme based

TABLE 1

List of disease complexes involving the reniform nematode, *R. reniformis*

Crop	Associated organism/s
Cotton	*Verticillium, Fusarium, R. solani, F. oxysporum, F. vasinfectum, Sclerotium rolfsi, V. dahliae, Glomus mosseae.*
Tomato	*G. mosseae, G. fasciculatum, M. incognita.*
Banana	*Radopholus similis, M. incognita.*
Brinjal	*Tylenchorhynchus brassicae.*
Black gram	*M. incognita.*
Black pepper	*M. incognita.*
Bushbean, cucumber, muskmelon	*Glomus mosseae.*
Grapes	*Fusarium solani, Agrobacterium tumefaciens*
Chillies	Marigold mottle virus.
Cantaloup	*Macrophomina phaseoli.*
Okra	*Rhizoctonia solani*, Yellow mosaic virus, *M. incognita.*

on the use of this volatile soil fumigant against *R. reniformis*. Initially, the four halide soil fumigants, D-D, EDB, DBCP and dichloropropene served as standard soil fumigants against the reniform nematode for the pineapple industry in Hawaii. With the withdrawal of most of these volatile halide soil fumigants from the commercial market in the United States of America, current interest in Hawaii has been now shifted to pre-planting application of dichloropropene at 224-336 l/ha and other non-volatile chemicals for the control of the nematode infected pineapple plants[25]. At present, non-volatile and contact chemicals including aldicarb, fensulfothion, carbofuran and other toxicants have been tested for control of the nematode attacking variety of crops lik grapes, banana, tobacco, cotton[42,64,66], soyban, cantaloupe[41], sweet potato[47] and potato[49]. The dosages used range from 2-16 kg a. i./ha, depending on the plant species involved.

Foliar application of systemic pesticide, oxamyl on tomato and cotton appears to be promising against the nematode. A single application of 600 and 2000 ppm of the chemical on tomato and cotton respectively appears to inhibit entry of the nematode into roots[47]. Significant reduction of soil population and yield increases in sweet potato are also reported with soil fumigants and several contact chemicals. Seed treatment of okra and pulses with carbofuran at the rate of 1.25 kg a.i./ha is effective against the entry of the parasite into root during the early stage of plant growth.

Resistance

In recent years, emphasis has been laid on the use of resistance as one of the most important components in the management of the nematode. Some important crops, in which resistance against the reniformis nematode has been located are listed in Table 2. Major efforts in this direction have been made on soybean. In the resistant cultivar 'Forrest' resistance was initially found to be governed by one major and minor gene pairs[48] but subsequently by two pairs of genes with unequal effect[49].

TABLE 2

List of resistant varieties/sources against reni m nematode

Crops/Plants	Resistant source/Variety
Beans	Jack bean ; Velvet bean
Coffee	*Coffea robusta* ; *C. canefora*
Cotton	Auburn 56, Plains, Delfos 425-920, Delfos 9169, Deltapine 6, Deltapine 6X,

(*Contd.*)

(*Contd.*)

	Delfos 6102, Cook 307-6X Delfos 9169, Lousiana Hybrid 33X 14-33, Louisiana 33, Roxie, Coker 100 wilt, Stoneville 2B strains
Cotton and its related species	*Gossypium longicahyse*, *G. barbadense*, *G. stocksil*, *G. somaelense*, *G. herbaceam*, *G. arboreum*, *G. hirsutum*, *Hibiscus* spp.
Cowpea	V-16
Lycopersicon pimpinellifolium	P-1-270453, PI-375937 Yellow strains
Tomato	Kalyanpur selection I, Kalyanpur selection III, LA 121 EC 118272, 11276, Peto 108, Peto 95, Nema 1400, ACE-55, Petopride
Mustard	Florida
Oats	Fulghum
Onion	Evergreen
Pepper and related crops	Sweet Ball, California wonder, *Capsicum annum* var. *fasciculatum*, *C. frutescens* var. *microcarpum*
Pigeon pea	UPAS-120, XPL-Prabhat, ICPL-87, ICPL-95, ICPL-96 and Pant A-10 ; 4658, 2387, 556-1, 4167, 3820-2 and Norman
Potato	La rouge, Red aboda B7 642-2, B 7154-8, B 7680-6, PI 37741
Sorghum	Sweet
Soybean	Peking Dyer, Centinal, Pickett, Pickett 71, Cluster, D 68-8, D-68-80, D 68-128, D 68-180, D 68-214, D 68-216 and D 68-221. TK 5, Strain 99, Strain 12, PI-200451 GC 40078-[illegible]-40, EGS [illegible], GC 40142-0-66, EGSY6 and VPL SY 2
Sweet potato	Gold rush

Other Methods

Amongst other methods, two years crop rotation using green sorghum or resistant soybean (Pickett) in place of cotton has been quite effective[50,51]. Hot water, organic amendments and antagonistic micro-organisms have been tried also with some degree of success. Hot water treatment is quite effective in eliminating nematode populations[52]. The mycorrhizal fungus, *Glomus fasciculatum*, has been demonstrated to increase resistance of tomato plants against nematode infection. The mycorrhizal fungus also affects adversely the life cycle of the nematode[53].

Rotylenchulus parvus (**Williams, 1960) Sher, 1961**

The nematode was originally described from sugarcane soil in Mauritius. It is also of common occurrence in corn and sugarcane fields of South Africa. It is also reported from soil around papaya in Nairobi, Kenya ; around corn, *Pennisetum typhoides*, *Crotolaria uncea* in Rhodesia and Turkish tobacco soil in Zambia. It has been recorded also from Puerto Rico and Imperial Valley, California (U.S.A.).

Host Range

Besides corn, sugarcane, black eye cowpea, common cowpea, sugarcane and red kidney bean, barley and Bermuda grass are the most favoured hosts[54].

Biology

Males are exceedingly rare. Apparently the nematode reproduces parthenogentically. The immature female is the only stage that infects the hosts and feeds on them. The anterior end of immature females enter the host to become a sedentary semi-endoparasite. The appearance of second-stage larvae is characterized by formation of a transversely striated cuticle beneath the smooth and loose skin of the first larval stage. The fully developed second-stage larvae emerge at about 11-13 days after first cleavage and undergo three superimposed moults without feeding. The fourth and final moult occurs at about 9-14 days after hatching. The life cycle from egg to egg takes about 27-36 days.

It can reproduce well at temperatures between 20-35°C. Under South African conditions, high clay content in soil favours nematode multiplication[55]. Despite the economic importance of sugarcane and corn cultivations in South Africa, very little information is available on the effect of hosts on the build-up of the population. Experiments carried out in fields of South Africa indicate that there is an increase of 3000 per cent in the number of the nematodes at the end of a 3 years corn monoculture[56].

Except for one brief report on histopathological changes in sugarcane roots, there is hardly any information available to judge pathogenic potentiality of this nematode. Cross sections of *R. parvus*-infected sugarcane roots show that the immature females penetrate host through cortical parenchymatous cells and come to rest in endodermis to establish feeding site[57]. The endodermal cell walls are deeply stained at the feeding site of the nematode. It is likely that feeding activity of this nematode may induce hypertrophy of endodermal and adjacent pericycle cells.

Rotylenchulus borealis Loof and Oostenbrink, 1962

Biology

The immature females are the only stage capable of infecting hosts. The life cycle from egg to egg stage is completed in 35-40 days at 20-25°C on corn roots[58].

Histopathological studies on *R. borealis* infected corn roots indicate that the immature females penetrate the cortex and establish feeding sites in endodermal cells which fuse with pericycle tissues to form a syncytium[57]. Besides the pericycle cells, phloem, protoxylem and vascular parenchymal cells are also involved in the formation of syncytium. Syncytical cells show very dense cytoplasm and accentuated wall fragmentation. In sweet potato roots, the nematode penetrates only 2-3 layers of cortical cells and does not reach stelar region. Permanent feeding sites are established in the cortical cells instead of endodermal cells which expand and fuse with adjacent cells to form syncytium. The syncytium subsequently expands into the stele by fusion with endodermis and pericycle cells. The syncytium appears as a funnel shaped structure with enlarged portion located in stele and narrow and elongated part in the cortical cells.

Like *R. reinformis*, *R. borealis* primarily feeds on endodermal cells, causing damage to pericycle cells. No information is available about its pathogenic potential on different crops.

Rotylenchulus macrodoratus **Dasgupta, Raski and Sher, 1968**

The nematode, commonly named as the Mediterranean reniform nematode, was described from soil around grape fields of Bari (Italy)[2]. It has been recorded also from France[59], Greece[60], Israel[61] and Malta[60]. It is of quite common occurrence in southern Itlay where it parasitizes roots of various fruit trees. The nematode usually prefers woody host plants like : *Ficus carica* (fig.), *Vitis* spp. (grapes), *Hedera halix* (ivy), *Laurel nobilis* (laurel), *Nerium oleandery* (oleander), *Olea europea* (olive), *Prunus amygdalus* (almond), *P. armeniaca* (apricot), *P. domestica* (plum), *Quercus calliprinos C. farnetto* (oak), *Pistacia vera* (pistachio), *Citrus sinensis* (orange), *Citrus* spp. and *Eriobotrys japonica* (loquat). It also parasitizes *Dianthus caryophyllus* (carnation), *Glycine hispida* (soybean), *Pholmis fruticosa* and *Dianthus barbatus*.

Biology

Like other species of the genus, *R. macrodoratus* is characterized in having infective stage limited to immature vermiform females only[62]. The immature females enter anteriorily to become sedentary semi-endoparasite. The mature females deposit eggs in gelatinous matrix. The life cycle, from egg to egg, is completed in 45-55 days, a period longer than known for any other species of *Rotylenchulus*. The egg masses usually contain 1-3 males and the ratio of males to immature females and larvae in a soil population is 1 : 10-15. No other information is available concerning various factors that influence the biology of the nematode.

Although histopathological studies for the nematode infested plant species like fig, grape, large flowered sweet william, olive, pistachio and soybean have heen carried out, there is very little work, so far, the assess pathogenic potentiality of the nematode[8,5].

Hyperplastic root reaction has been reported in late flowering sweet william and pistachio. In pistachio, galls usually occur along the root axis, with 2-3 females attached to swollen portion of root.

An unique feature is the formation of an uninucleate giant cell originating usually from endodermis[63]. This type of host reaction is not encountered in any of the other species of the genus. Light and electron microscopic studies on *R. macrodoratus*-infected soybean roots have revealed that an uninucleate giant cell is usually initiated in the pericycle or endodermal tissues and is expanded into stele. Occasionally, the nematode penetrates the central stele initiating giant cell formation which subsequently expands to the endodermis. Enlargement of giant cell results in compression and distortion of adjacent parenchymal cells. Increased cytoplasmic density, organelle proliferation, nuclear hypertrophy, lobing and cell wall ingrowth are the intracellular changes associated with progressive expansion of giant cell. The single nucleus of the giant cell is maintained throughout the development of the enlarged cell. Thus, giant cell in this type of host-parasite reaction appears to be the result of expansion of a single cell on which the nematode sustains.

On olive roots, the nematode feeds on both cortex and endodermis. The feeding site consists of an uninucleate giant cell with dense cytoplasm, enlarged nucleus and nucleolus. In the close proximity of nematode head, the cell wall produces ingrowth into which the stylet is directly penetrated. The giant cell or 'nurse cell' occupies stele area, if it is of endodermal origin. In contrast, it occupies cortex and stele area, if the giant cell originates from a cortical cell. Vascular tissue, especially the xylem, is compressed and crushed due to expansion of this enlarged cell. Epidermal and cortical cells, adjacent to nematode body, appears to be necrotic with thick wall.

In pistacio, there is hyperplastic reaction of the secondary vascular parenchyma cells surrounding the uninucleate giant cell[64]. Hyperplasia of tissue and enlargement of giant cell result in assymetric structure of root with central cylinder shifted to one side of root.

Rotylenchulus sacchari Van der Berg and Spaull, 1981

The potentiality of *R. sacchari*, like many other species of the genus, as a plant pathogen, is yet unknown. It appears to be confined to sandy clay loam soil in South Africa.

REFERENCES

1. Linford, M.B. and Oliveira, J.M., *Rotylenchulus reniformis* nov. gen., n.sp., a nematode parasite of roots. *Proc. Helminth. Soc. Wash.*, 7, 34-42, 1940.

2. Dasgupta, D.R., Raski, D.J. and Sher, S.A., A revision of the genus *Rotylenchulus* Linford & Oliveira (Nematoda : Tylenchidae) *Proc. Helminth. Soc. Wash.*, 35, 189-192, 1968.

3. Swarup, G., Nath, R.P. and Sethi, C.L., The plant parasitic nematode genus *Rotylenchulus* in India. *Indian phytopath.*, 20. 118-123.

4. Smith, A.L. and Taylor, A.L., Nematode distribution in 1940 regional cotton wilt plots. *Phytopathology*, 31, 777 (abstr.), 1941.

5. Norton, D.C., Plant parasitic nematodes in Texas. *Texas Agric. Expt. Sta.*, Misc. Publ., 321, pp. 1959.

6. Minton, N.A. and Hopper, B.E., The reniform and sting nematodes in Alabama. *Plant Dis. Reptr.*, 43, 47 p., 1959.

7. Fassuliotis, G. and Rau, G.J., The reniform nematode in South Carolina. *Plant Dis. Reptr.*, 51, 557, 1967.

8. Holdeman, Q., Cordas, D., Watson, T., Matsumoto, R. and Siddioni, I., Fact finding study. The reniform nematode, *Rotylenchulus reniformis.* Calif. Dept. Food & Agric., Division of Plant Industry Study Team 1973-1974, 1977.

9. Heald, C.M. and Thames, W.H., The reniform nematode, *Rotylenchulus reniformis*, 139-143. In : Nematology in the southern region of the United States So. Coopt. Series Bull. No. 276, 206, 1982.

10. Linford, M.B. and Yap, F., Some host plants of the reniform nematode in Hawaii. *Proc. Helminth. Soc. Wash.*, 7, 42-44, 1940.

11. Birchfield, W. and Brister, I.R., New host and non-hosts of reniform nematodes. *Plant Dis. Reptr.*, 46, 683-685, 1962.

12. Peacock, E.C., The reniform nematode in Gold Coast. *Nematologica*, 1, 307-310, 1956.

13. Birchfield, W., Host-parasite relation of *Rotylenchulus* on *Gossypium hirsutum*. *Phytopathology*, 52, 862-865, 1962.

14. Sivakumar, C.V. and Seshadri, A.R., Life history of the reniform nematode, *Rotylenchulus reniformis* Linford & Oliveira, 1940. *Indian J. Nematol.*, 1, 7-20, 1971.

15. Dasgupta, D.R. and Seshadri, A.R., Reproduction, hybridization and host adaptation in physiological races of the reniform nematode, *Rotylenchulus reniformis*. *Indian J. Nematol.*, 1, 128-144, 1971.

16. Nakasono, K., Sexual attraction of *Rotylenchulus reniformis* females and some differences in the attractiveness between amphimictic and parthenogenetic populations (Nematoda : Nacobbidae). L.C. in *Helminthol. Abstr.*, 47, 31 (Abstr.), 1978.

17. Singh, R.V. and Khera, S., Culturing and life history studies on *Rotylenchulus reniformis* Linford & Oliveira on brinjal (Nematoda). *Bull. Zool. Surv. India*, 1, 115-128, 1978.

18. Rodriguez-Fuents, M.E., Flerdona, L.C., *In vitro* studies on the influence of humidity on *Rotylenchulus reniformis*. *Central Agricola*, 8, 39-42, 1981.

19. Khan, F.A. and Khan, A.M., Studies on the reniform nematode, *Rotylenchulus reniformis*. I. Host range and population changes. *Indian J. Nematol.*, 3, 24-30, 1974.

20. Tasai, B.Y. and Abi, W.J., Anhydrobiosis of reniform nematodes : Survival and coiling. *J. Nematol.*, 11, 36, 1979.

21. Birchfield, W. and Martin, W.J., Reniform nematode survival in air-dried soil. *Phytopathology*, 57, 804 (Abstr.), 1967.

22. Apt, W.J., Survival of reniform nematodes in desiccated soils. *J. Nematol.*, 8, 273 (Abstr.), 1976.

23. Heald, C.M,, Pathogenecity and histopathology of *Rotylenchulus reniformis* infecting cantaloup. *J. Nematol.*, 7, 149-152, 1975.

24. Rebois, R.V., Effect of soil temperature on infectivity and development of *Rotylenchulus reniformis* on resistant and susceptible soybean *Glycin max. J. Nematol.*, 5, 10-13, 1973.

25. Rohrbach. K.G. and Apt, W.J., Nematode and disease problems of pineapple. *Plant Disease*, 70, 81-87, 1986.

26. Heald, C.M. and Heilman, M.D., Interaction of *Rotylenchulus reniformis*, soil salinity and cotton. *J. Nematol.*, 3, 179-182, 1971.

27. Ayala, A., Pathogenecity of reniform nematode on various hosts. *J. Agr. Univ.* Puerto Rico, 46, 63-82, 1963.

28. Varaprasad, K.S., *Rotylenchulus reniformis* Linford & Olivera, 1940 : A comparative account of systematics, biology and management, pp. 194-92. In : Plant Parasitic Nematodes of India : Problems and Progress, IARI Publication, 497 pp. 1986.

29. Dasgupta, D.R. and Seshadri, A.R., Races of the reniform nematode, *Rotylenchulus reniformis* Linford & Oliveira, 1940. *Indian J. Nematol.*, 1, 21-24, 1971.

30. Heald, C.M., Effect of reniform nematode on cantaloup yields. *Plant Dis. Reptr.*, 64, 282-283, 1980.

31. Jones, J.E., Newsom, L.D. and Finley, E.J., Effect of reniform nematode on yield, plant character and fiber properties of upland cotton. *Agron. J.*, 51, 353-356, 1958.

32. Singh. N.D., Studies on selected hosts of *Rotylenchulus reniformis* and its pathogenecity to soybean (*Glycine max*). *Nematropica*, 5, 46-51, 1975.

33. Castillo, M.B., Arceo, M.B. and Litsinger, J.A., Population dynamics of plant parasitic nematodes. 1. *Rotylenchulus reniformis* and its effect on yields of field legumes. *Phil. Agr.*, 61, 238 252, Dec. 1977-1978.

34. Rebois, R.V., The effect of *Rotylenchulus reniformis* inoculum level on yield, nitrogen, potassium, phosphorus and amino acids of seed of resistant and susceptible soybean (*Glycine max*). *J. Nematol*, 3, 326-327, (Abstr.), 1971

35. Sivakumar, C.V. and Seshadri, A.R., Pathogenecity of the reniform nematode, *Rotylenchulus reniformis* to castor. *Indian J Nematol.*, 1, 227-236, 1971.

36. Rebois, R.V. and Eldrige, B.J. and Webb, R.E., *Rotylenchulus reniformis* parasitism and its effect on yields. *Plant Dis. Reptr.* 62, 520-523, 1978.

37. Sivakumar, C.V. and Seshadri, A.R., Histopathology of infection by the reniform nematode, *Rotylenchulus reniformis* Linford & Oliveira, 1940 on castor, tomato and papaya. *J. Nematol.*, 2, 175-181, 1972.

38. Rebois, R.V., Epps, J.M. and Hartwig, E.E., Correlation of resistance in soybeans to *Heterodera glycines* and *Rotylenchulus reniformis*. *Phytopathology*, 60, 695-700, 1970.

39. Rebois, R.V., Ultrastructure of a feeding peg and tube associated with *Rotylenchulus reniformis* in cotton. *Nematologica*, 26, 326-405, 1980.

40. Rebois, R.V., Madden, P.A. and Eldrige, B.J., Some ultrastructural changes induced in resistant and susceptible soybean roots following infection by *Rotylenchulus reniformis*. *J. Nematol.*, 7, 122-139, 1975.

41. Cohn, E., Histology of the feeding site of *Rotylenchulus reniformis*. *Nematologica*, 19, 455-458, 1973.

42. Razak, A.R. and Evans, A.A., An intracellular tube associated

with feeding by *Rotylenchnlus reniformis* on cowpea roots. *Nematologica*, 22, 182-184, 1976.

43. Heald, C.M., Pathogenecity and histopathology of *Rotylenchulus reniformis* infecting cantaloupe. *J. Nematol.*, 7, 149-152, 1975.

44. Robinson, A.F. and Orr, C.C., Histopathology of *Rotylenchulus reniformis* on sunflower. *J. Nematol.*, 12, 84-85 (Abstr.), 1980.

45. Neal, D.C., *Fusarium* wilt and fiber properties of some new cotton hybrids and selections. *Phytopathology*, 43, 295 (Abstr.) 1953.

46. Neal, D.C., The reniform nematode and its relationship to the incidence of *Fusarium* wilt of cotton at Baton Rouge., Louisiana. *Phytopathology*, 44, 447-450, 1954.

47. Rich, J.R. and Bird, G.W., Inhibition of *Rotylenchulus reniformis* penetrations of tomato and cotton roots with foliar application oxamyl. *J. Nematol.*, 5, 221-224, 1980.

48. Gillman, D.F., Marshall, J.G., Rabb, J.H., Laurence, R.M., Habetz, R., Allen, M., Boques, D. and Bartleson, J.L., Performance traits with soybean varieties in Louisiana. La. Agric. Expt. Sta. Agron. Res. Rept., 49, 2400, 1977.

49. Harvillie, D.G., Green, A. and Birchfield, W. Genetic resistance to reniform nematodes in soybeans. *Plant Dis. Rep'r*,, 69, 587-589, 1985.

50. Thames, W.H. and Heald, C.M., Chemical control of *Rotylenchulus reniformis* on cotton. *Plant Dis. Reptr.*, 58, 337 341, 1974.

51. Gillman, D.F., Jones, J.E., Williams, C. and Birchfield, W., Cotton-soybean rotation for control of reniform nematodes. *La. Agric.*, 21, 10-11, 1978.

52. Martin, W.J., Elimination of root-knot and reniform nematodes and scurf infections from root lets of sweet potato plants by hot water treatment. *Plant Dis. Reptr.*, 54, 1056-1058, 1970.

53. Sitaramhaiah, K. and Sikora, R.A., Effect of the mycorrhizal fungus *Glomus fasciculatus* on the host-parasite relationship of *Rotylenchulus reniformis* in tomato. *Nematologica*, 28, 412-419, 1982.

54. Dasgupta, D.R. and Raski, D.J., The biology of *Rotylenchulus parvus*. *Nematologica*. 14, 429-440, 1968.

55. Berg, Van Der, E. and Spaull, V.W., *Rotylenchulus* species found associated with sugarcane in south Africa with a description of *R. sacchari* sp.n. (Nematoda : Rotylenchulinae). *Phytophylactia*, 13, 43-47, 1981.

56. Heyns, J., *Rotylenchulus parvus*. C.I.H. descriptions of plant parasitic nematodes. Set. 6, No. 83, 1976.

57. Volvas, N., Cohn, E. and Inserra, R.N., Histological changes induced by *Rotylenchulus borealis* on corn and sweet potato and *R. parvus* on sugarcane. *Revue Nematol.*, 8, 125-130, 1985.

58. Volvas, N. and Inserra, R.N., Biological relationship of *Rotylenchulus borealis* on several plant cultivars. *J. Nematol.*, 14, 373-377, 1988.

59. Scotto La Massese, C., Nouvel hote et nouvelle localisation de *Rotylenchulus macrodoratus* Dasgupta, Raski & Sher, 1968. *Nematol. Medit.*, 1, 55-56, 1973.

60. Volvas, N. and Lamberti, F., Nuoviospiti naturali di *Rotylenchulus macrodoratus* Dasgupta, Raski & Sher, 1968. Malta region. *Nematol. Medit.*, 2, 177-179, 1974.

61. Cohn., E. and Mondechai, M., Ultrastructure of hypertrophied cell induced in soybean roots by *Rotylenchulus macrodoratus*. XIII. *Int. Nematol. Symp.*, Dublin, Ireland (Sept. 5-11), pp. 17-18, 1976.

62. Inserra, R.N. and Volvas, N., The biology of *Rotylenchulus macrodoratus*. *J. Nematol.*, 12, 97-102, 1980.

63. Cohn, E. & Mordechai, M., Uninucleate giant cell induced in soybean by the nematode *Rotylenchulus macrodoratus*. *Phytoparasitica*, 5, 85-93, 1977.

64. Volvas, N., Gall formation on *Pistacia vera* by *Rotylenchulus macrodoratus*. *J. Nematol.*, 15, 148-150, 1985.

12

Tylenchulus semipenetrans Cobb, 1913

Citrus species occupy an important position amongst the fruit crops, particularly in the tropical and subtropical areas of the world. The area under commercial orchards of different citrus species, for consumption of fresh fruits and processing industries, has expanded considerably during the last three to four decades. However, there has not been any significant increase in total area-production ratio. One of the important constraints are nematodes, particularly the citrus nematode. Even though, the first record of association of a plant parasitic nematode was that of root-knot nematodes, as early as 1889[1], recognition of nematodes as important limiting factors in citriculture came only after the discovery of the association of the citrus nematode in 1912[2] in the State of California, U.S.A. It was subsequently found to be associated with unthrifty citrus orchards in other citrus growing areas of U.S.A. and also in Malta, Spain, Israel, Australia and South America. This is now accepted as one of the most important nematode pathogens of citrus and which is most often associated with the die back complex—still an intensively discussed problem. The nematode is now virtually recorded from all citrus orchards of the world and is considered to cause 'virtually immeasureable loss' to the citrus industry[3].

Morphology and Biology

A most comprehensive account has been given by Van

Gundy[4] in 1958 and there has not been any significant contribution on these aspects since then.

Females are saccate with short blunt tail. Most commonly, these are found attached to the short and thick roots. At the point of attachment, the exposed posterior part of the female is covered with soil particles. The eggs are laid in a mucoid membranous structure surrounding the vaginal area and are protected from the covering of soil particles also. Under laboratory conditions, the second-stage larvae emerge from the eggs after 12-14 days. Two different larval sizes, short and long, emerge from the eggs. The short larvae have the ability to develop into males without feeding but the larvae (long) which are destined to be females require parasitisation of roots for further development. Once these larvae have penetrated the roots, further development takes place while they are attached with the roots. The developing females gradually become swollen into saccate-like structure and secrete a gelatinous matrix through the excretory pore (Fig. 1). The matrix alongwith the adhering soil and organic matter give protection to the nematode from natural predators[5]. The life cycle (egg to egg) is completed between 4-8 weeks[6-8].

Hosts

Mainly confined to the members of the family Rutaceae. A total of 75 rutaceous species, 50 of which comprise of *Citrus* species or citrus hybrids and 9 non-rutaceous species have been recorded as hosts of the nematode[9,10]. Important commercial species like grapefruit (*Citrus paradisi*), sweet orange (*C. sinensis*) and sour orange (*C. aurantium*) have been recorded as susceptible hosts[11]. So far, there are no reports in the literature indicating any resistance in any of the *Citrus* species. However, the relative host status may differ, with some of these species highly susceptible than others.

Some unusual recorded hosts are : *Olea europaea*, *Vitis vinifera*, *Syringia vulgaris*, *Andropogon rhizomatus*, *Calodendron capense*, *Eriobotrys japonica*, *Mikania batatifolia*, *Prunus communis*, *Diospyros lotus*, and *Garcinia mangostana*. All these are non-rutaceous hosts[12].

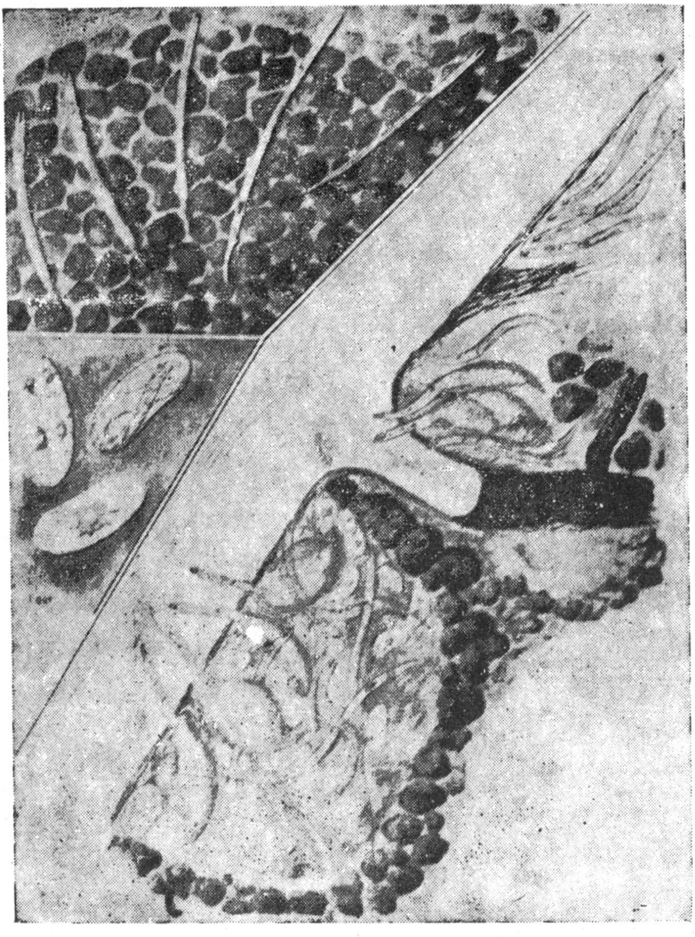

Fig. 1. Life cycle of the citrus nematode.
(After drawings by Charles. Papp., University of California, USA).

Symptoms

Generally, the nematode is host specific, confined to *Citrus* species and associated with varied type of disease symptoms, particularly 'slow decline or slow die-back or die-back'. General decline of infested trees, defoliation of shoots, smalling and yellowing of leaves and fruits, sign of starvation in leaves (Fig. 2*a* and *b*), necrotic stripes on roots, sloughing off of roots ultimately leading to

Fig. 2 (*a*). General decline of infested trees :
Declining tree.

Fig. 2 (*b*). General decline of infested trees :
A citrus orchard with several declined trees.

Fig. 3(*a*). Citrus root damage caused by *Tylenchulus semipenetrans* :
Left—Healthy roots ; Right—Roots after inoculations with citrus nematode.

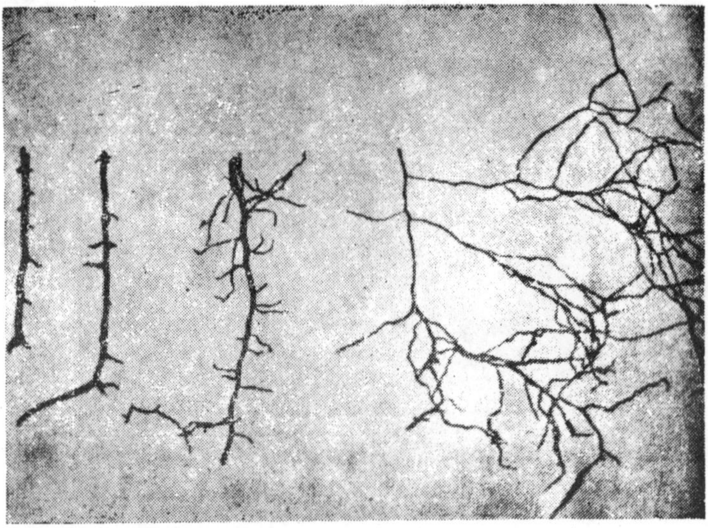

Fig. 3 (*b*). Citrus root damage caused by *Tylenchulus semipenetrans* : Left—Diseased roots exhibiting lack of feeder roots ; Right—Healthy roots.

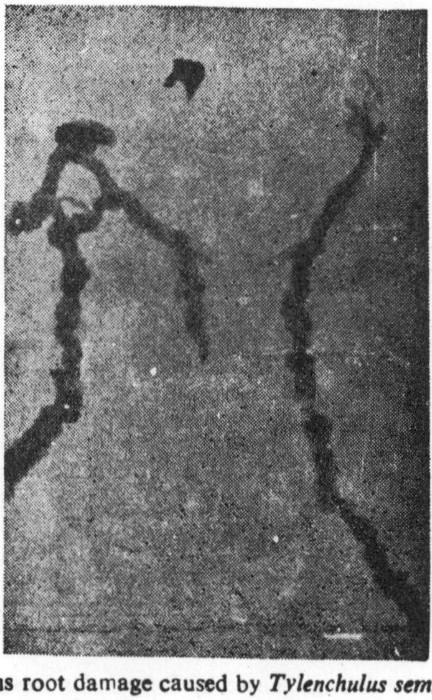

Fig. 3 (*c*). Citrus root damage caused by *Tylenchulus semipenetrans* : Diseased roots exhibiting sloughing.

root decay are some of broad effects (Fig. 3 *a*, *b* and *c*) recorded under pathogenicity tests[13-17]. Pathogenicity and manifestation of symptoms are closely related to nematode density, with the trees deteriorating when the level becomes critical[18]. It has been suggested also that infestation leads to root damage, impairment of normal growth function and uptake of nutrients[19]. Secondary invasion of soil fungi particularly *Fusarium solani* and *F. oxysporum* aggravate disease manifestation, especially in the temperature range of 25-30°C[20,21].

The citrus trees exhibiting early symptoms of decline are recorded to harbour large populations of the citrus nematode as against low numbers of the nematode in declined or in advanced declined trees. It has been suggested that with high nematode population, there is a rapid deterioration of the root system which ultimately affects the nematode population because of non-availability of feeder roots. Such trees decline within 3-5 years[22,23].

There are variable reports about the extent of loss to the citrus industry by this nematode, the range being from 10 to 60 per cent with respect to growth of seedlings and 10 to 20 per cent for yield loss[4,23-26]. Soil pH, moisture and clay content of the soil are important determinants in the extent of damage to the trees. Maximum damage has been recorded at 5.0 to 7.5 pH, with moderate amount of sodium and excess calcium carbonate at low level of soil phosphorus[27].

The nematode is reported to cause damage by mechanical injury to roots, interfering with normal root functions and injection of some toxic factor in the roots[14]. The nematode is known to alter the semi-permeable nature of the cells, permitting roots to absorb large amounts of some elements. Irrigation with salty water causes accumulation of salts in the trees resulting in pronounced nematode damage symptoms[27,28]. Slow invasion of citrus roots by nematode larvae as well as relatively long life cycle, even under optimal conditions, protect the trees in orchards, atleast for the first few years[15]. Non-infested root stocks, planted in infested soils may show damage symptoms in five years[29]. However, plantings of infected root stocks in virgin soils may permit much longer span of relief from nematode damage. Under low nematode population conditions, root recovery is possible. The damage symptoms of citrus decline become evident only when the delicate equilibrium between root recovery and root infection gets upset with more and more feeder roots getting infected. Usually the peak or 'ceiling' populations are achieved after 12-17 years. In comparison to young apparently healthy trees, trees in advanced stages of citrus decline support very low nematode population[30,31].

There is hardly any informations available about the involvement of other soil micro-organisms with the citrus nematode in disease syndrome. The only mention is that of *Fusarium solani* which has been found to cause greater reduction growth in presence of the nematode than when alone[32,33].

Predominantly, the spread is through the movement of infested rootstocks to new areas. Movement of soil accounts for only local and limited distribution. Within the orchards, irrigation water is a prime source in the movement of nematode populations. On their own, the nematode movement is very limited—not more than 53 cm in two years time[34].

The juveniles can survive for fairly long periods of time even in the absence of host. According to some estimates, these can survive without host for 9 years[28] and can withstand as high a temperature as 45°C for several hours[35]. The juveniles can survive also in water at 10°C for 24 months[36].

Management

Perennial nature of the tree, deep as well as widespread root system makes the control of the nematode a somewhat difficult proposition. As with many other nematodes, once this gets established in a new site, it becomes virtually impossible to eradicate it. Therefore, a judicious mixture of different approaches has to be followed to devise an economically viable management strategy. Obviously, of foremost importance is regulation of the movement of the infested rootstocks in uninfested areas. Raising citrus rootstocks in virgin uninfested nurseries will go a long way in limiting the build up of the nematode population. Further, the seedlings to be transplanted can be given hot water treatment at 45°C for 25 minutes[37] or treated with aqueous solution of some systemic organophosphates[38].

By far, the most effective results have been obtained through the use of chemicals, which often may not be economical. Chemicals can be applied either as pre-plant or post-plant treatment. Methyl bromide and chloropicrin have been used in the past but these are quite expensive. Economically, it is feasible to treat nurseries before planting nursery stocks. The biggest disadvantage associated with pre-planting chemicals is their phytotoxicity. Consequently, some interval has to be given between planting and treatment otherwise temporary inhibition in root growth may take place[39].

DBCP, as a post-plant treatment, highlighted the potential of the trees under reduced nematode population levels. In fact, it became quite popular with the citrus growers as a must soil drench treatment[40,41]. However this chemical has become largely unavailable and systemic organophosphates and carbamates and some other groups are being used instead. Some of these have been proved to be quite effective in reducing nematode populations as well as increasing yields of fruits[42–45]. However, none of these have resulted in the same yield increases as was obtained with DBCP application. One of the reasons for DBCP popularity, besides its effectiveness, was its easy miscibility with water. The latter aspect was somewhat achieved with phenamiphos 40 per cent E.C. formulation. At a dose of 10 ml/m^2, it resulted in increase of orange yields by 6.1 to 7.5 tons/ha, besides improving the uptake efficiency of K, Ca and Cu[46]. Applications of carbofuran, aldoxycarb and oxamyl at 1.1 kg a.i./ha by low pressure drip irrigation system have proved also beneficial to citrus trees[47]. Some of these chemicals can be used effectively as soil treatment also.

The other approaches in management practices include soil amendment, use of inter crops, biological control and use of resistant rootstocks. Application of cowdung manure diluted with water or use of sewage water for irrigation or application of refuse compost have yielded good results with respect to decreasing nematode populations and increasing fruit yields[18,34,48]. A possible breakthrough in nematode management is being projected with the use of organisms like *Paecilomyces lilacinus* which has proved to be effective in reducing the nematode population as well as increasing the fruit size[49].

Identification of resistant sources in *Citrus* species and varieties has been moderately successful, with most important source as selections of trifoliate orange (*Poncirus trifoliata*). The intergeneric hybrids formed by crossing *P. trifoliata* with susceptible *Citrus* species have been recorded to exhibit resistance[50]. Resistance breaking biotypes are, however, also recorded in the nematode populations.

REFERENCES

1. Neal, J.C., Root-knot disease of peach, orange and other plants in Florida, due to the work of *Anguillula. USDA Bur. Entomol. Bull.*, No. 20, 1-31, 1889.
2. Thomas, E.E., A preliminary report of a nematode observed on citrus roots and its possible relation with the mottled appearance of citrus trees. *Calif. Agri. Exp. Sta. Circ.*, 85 1913.
3. Ducharme, E.P., In : Nematodes of Tropical Crops. J.E. Peachey (ed.). *Tech. Comm.* No. 40, CAB, 225-237, 1969.
4. Van Gundy, S.D., The life history of the citrus nematode, *Tylenchulus semipenetrans. Nematologica*, 3, 283-294, 1958.
5. Maggenti, A.R., The production of the gelatinous matrix and its taxonomic significance in *Tylenchulus. Proc. helm. Soc. Wash.*, 29, 139-144, 1962.
6. Cohn, E., The development of the citrus nematode on some of its hosts. *Nematologica*, 11, 593-600, 1965.
7. Macaron, J., Contribution to the study of the plant-parasitic nematode, *Tylenchulus semipenetrans*. Ph.D. Thesis, *Univ. Sci. & Technol.*, Languedoc, Monthpellier, 1972.
8. O'Bannon, J.H., Reynolds, H.W. and Leathers, C.R., Effects of temperature on penetration, development and reproduction of *Tylenchulus semipenetrans. Nematologica*, 12, 483-487, 1966
9. Vilardebo, A. and Luc, M., Slow decline of citrus caused the nematode *Tylenchulus semipenetrans. Fruits* (Paris), 455-465, 1961.

10. Cohn, E. and Milne, D.L., New host plants of the citrus nematode from South Africa. *Plant Dis. Reptr.*, 61, 466-467, 1977.

11. Cobb, N.A., Citrus-root nematode. *J. Agric. Res.*, 2, 217-230. 1914.

12. Chawla, M.L., Samathadan G.J. and Sharma, S.B., Occurrence of citrus nematode (*Tylenchulus semipenetrans*) on roots of mangosteen (*Garcinia mangostna*). *Indian J. Nematol.*, 10, 240-242, 1980.

13. Thomas, E.E., The citrus nematode, *Tylenchulus semipenetrans*. *Univ. Calif. Agric. Exp. Sta., Coll. Agric. Tech. Paper* 2, 1923.

14. Baines, R.C. and Clarke, O.F., Some effects of the citrus root nematode on the growth of orange and lemon trees. *Phytopathology*, 42, 1952.

15. Cohn, E., Host response to the penetration of the citrus nematode, *Tylenchulus semipenetrans*. *Nematologica*, 10, 77-78, 1964.

16. Natur, R.M., El Haidari, H. and Mohammed, J., Citrus root nematode in Iraq. *Pl. Dis.. Reptr.*, 49, 792, 1965.

17. El-Haidari, H.S., Citrus-root nematode in Lebanon. *Pl. Dis. Reptr.*, 50, 960, 1966.

18. Cohn, E., Minz, G. and Monselise, S.P., The distribution, ecology and pathogenecity of the citrus nematode in Israel. *Israel J. Agric. Res.*, 15, 187-200, 1965.

19. Bindra, O.S. and Chhabra, H.K., Control of citrus nematode may halt decline of your orchard. *Prog. Fmg.*, 5, 17-18, 1968.

20. Van Gundy, S.D. and Tsao, R.H., Growth reduction of citrus seedings by *Fusarium solani* as influenced by the citrus nematode and other soil factors. *Phytopathology*, 53, 488-489, 1963.

21. O'Bannon, J.H., Leathers, C.R. and Reynolds, H.W., Interactions of *Tylenchulus semipenetrans* and *Fusarium* species on

rough lemon (*Citrus limon*). *Phytopathology*, 57, 414-417, 1967.

22. Reynolds, H.W. and O'Bannon, J.H., Decline of grape fruit trees in relation to citrus nematode populations and tree recovery after chemical treatment. *Phytopathology*, 53, 1011-1015, 1963.

23. Chona. B.L., Sethi, C.L., Swarup, G. and Gill, J.S., Citrus nematode (*Tylanchulus semipenetrans*) associated with die-back disease of citrus trees in India. *Indian J. Hort.* 22, 370-373, 1966.

24. Baines, R.C. and Clarke, O.F., Citrus nematode. *Calif. Agric.*, 6, 9, 1952.

25. Baines, R.C., Foote, F.J. and Martin, J.P., Fumigate soil before replanting citrus for control of citrus nematode. *Calif. Agric. Exp. Sta.*, Leaflet, 91, 1957.

26. Baines, R.C., Small R.H.. Dewolfe, T.A., Martin, J.P. and Stolzy, L.H., Control of the citrus nematode and *Phytophthora* species by vapam. *Plant Dis. Reptr.*, 41, 405-414, 1957.

27. Van Gundy, S.D. and Martin, J.P., Influence of *Tylenchulus semipenetrans* on the growth and chemical composition of sweet orange seedlings in soils of various exchangeable cation ratios. *Phytopathology*, 51, 146-151, 1961.

28. Baines, R.C., Martin, J.P., Dewolfe, T.A., Boswell, S.B. and Garber, M.J., Effect of high doses of D-D on soil organisms and the growth and yield of lemon trees. *Phytopathology*, 52, 723, 1962.

29. O'Bannon, J.H. and Tarjan, A.C., Preplant fumigation for citrus nematode control in Florida. *J. Nematol.*, 5, 88-95, 1973.

30. Swarup, G., Sethi, S.L. and Gill, J.S., Some records of plant parasitic nematodes in India. *Curr. Sci.*, 33, 593, 1964.

31. Ahmed, N.S., M.Sc. (Ag.) Thesis. S.V. Agricultural College, (A.P.A.U.). Triupati, 83 pp.. 1985.

32. Thomas, E.E., The citrus nematode, *Tylenchulus semipenetrans. Univ. Calif. Agric. Exp. Sta. Coll. Agri. Tech. Paper* 2, 1923.

33. Van Gundy, S.D. and Tsao, P.H., Growth reduction of citrus seedlings by *Fusarium solani* as influenced by the citrus nematode and other soil factors. *Phytopathology*, 53, 488-489, 1963.

34. Tarjan, A.C., Migration of three pathogenic citrus nematodes through two Florida soils. *Soil Crop Sci. Soc. Fla. Proc.*, 31, 253-255, 1971.

35. Feldmesser, J. and Rebois, R.V., Effects of soil temperature and moisture on *Tylenchulus semipenetrans. Phytopathology*, 53, 875, 1963.

36. Cohn, E., Observations on the survival of free-living stages of the citrus nematode *Nematologica*, 12, 321-327, 1966.

37. Baines, R.C., Nematodes on citrus. *Calif. Agric.*, 4, 7, 1950.

38. O'Bannon, J.H. and Taylor, A.L., Control of nematodes on citrus seedlings by chemical bare-root dip, *Plant Dis. Reptr.*, 51, 99.-998, 1967.

39. Cohn, E., Feder, W.A. and Mordechai, M., The growth responses of citrus to nematicide treatments. *Israel J. Agric. Res.*, 8, 19 24, 1968.

40. Baines, R.C., Bitters, W.P. and Clarke, O.F., Susceptibility of some species and varieties of citrus and some other rutaceous plants to the citrus nematode. *Plant Dis. Reptr.*, 44, 281-285, 1960.

41. Cohn, E. and Minz, G., Application of nematicides in established orchards for controlling the citrus nematode, *Tylenchulus semipenetrans. Phytopathol. Medit.*, 4, 17-20.

42. Baines, R.C. and Small, R.H., Evidence of modes of action of oxamyl nematicides on *Tylenchulus semipenetrans. J. Nematol.*, 6, 135, 1974.

43. Baines, R.C. and Small, R.H., Effects of oxamyl on the citrus nematode, *Tylenchulus semipenetrans* and on infection of sweet orange. *J. Nematol.*, 8, 122-130, 1976.

44. Abu-Elamayem, M.M., Rabie, R.S., El-Sabae, A.H. and Tentawy, G.A., Foliar application of certain nematicides for the control of root-knot and citrus nematodes. Beitrage Zur Tropisechen. Lendcuirtschaft und Veterinarmdizin 19, 229-234, 1981.

45. Hamid, G.A., Van Gundy, S.D., Lovatt, C.J., Citrus nematode alters carbohydrate partitioning in the 'Washington' Navel orange. *J. Amer. Soc. Hort. Sci.*, 110, 642-646, 1985.

46. Milne, D.L. and Willers, P., Yield and nutritional responses due to phenamiphos treatment of citrus infested with citrus nematode. *Citrus and Subtropical Frui' J.*, No. 547, 22-24, 1979.

47. Van Gundy, S.D. and Garabedian, S., Application of nematicides through drip irrigation systems. *Medtd de Facult.* Landbou. Rajksuniversiteit, Gent 49, 629-634, 1984.

48. Tichinova, L.V. In : Eelworms of Agriculture Crops in them Vzbek. SSR" Tashkent : Akad. Nauk, Vzebksoi SSR, 101-131, 1957.

49. Jatala, P., In : An advanced treatise on *Meloidogyne* vol. I, Biology and Control (Eds.) Sasser, J.N. and Carter, C.C. North Car. S V. Graphics, pp. 303-308, 1985.

50. McCarty, D D., Bitters, W.P. and Van Gundy, S.D., Susceptibility of rootstocks to the citrus nematode. *Calif. citrogr.*, 64, 129-144, 1979.

13

Genus Hirschmanniella Luc and Goodey, 1964

There are twentytwo recognised species in the genus and the best known species is *H. oryzae* (Van Breda de Haan, 1902) Luc and Goodey, 1964. It is one of the most important migratory endoparasites, often an important limiting factor in rice cultivation, recorded in the begining of this century from Java (Indonesia)[1] under the name *Tylenchus oryzae*. Almost after three decades, a detailed description along with illustrations was published on the basis of specimens collected from Java[2,3]. It was implicated as the cause of mentek/omomentek disease of rice in Indonesia. However, the precise role of the nematode in the disease has eluded clarifications, so far[4,5]. Inspite of the fact that no definite evidences could be found about its role in the 'mentek' disease, its pathogenic potentiality has been demonstrated on rice in several countries.

Two other species of some economic importance are : *H. imamuri* Sher, 1968 and *H. spinicaudata* (Schuurman Stekohoven, 1944) Luc and Goodey, 1964.

Morphology

Generally, the nematodes are usually long and slender, measuring over 1 mm in length. Cephalic region is continuous, anteriorly flattened or hemispherical with heavily sclerotized framework. Stylet is massive, 15-46 μm long with large strong knobs Esophageal glands are elongated, lying ventral to intestine, with asymmetrical sub-ventral glands which are longer than the dorsal one. The females are with 2 gonads, outstretched and opposite. Tails are similar in both the sexes, usually with a terminal mucro.

Distribution and Host Range

The nematodes are cosmopolitan in distribution. *H. oryzae* occurs in Asia, Africa, South and North America. Another species of economic importance, *H. spinicaudata* is prevalent in African and Asian continents. *H. oryzae* and *H. spinicaudata*, in West Africa, show definite geographical distribution-pattern. *H. spinicaudata* is highly prevalent in humid region like Ivory Coast, *H. oryzae* mostly in Senegal and mixture of both species in intermediate areas like Gambia[6]. The third species of economic importance, *H. imamuri* is prevalent in Japan. Most of the described species appear to be confined either in Asian or African continent. Table 1 illustrates geographical distribution of some nominal species. Host range has been given in Table 2.

Biology

Because of its economic importance, *H. oryzae* has been investigated in details in comparison to other species[3,7]. The oviposition begins within a few days after penetration, followed by appearance of the first-stage larvae, lacking stylet but with discernible esophagus, in egg shell, after 96 hrs. Fully developed second-stage larvae resemble adults except for shorter body size, stylet and oesophageal glands. The second stage larvae undergo two more moults to become third- and fourth-stage larvae. The second and third moults occur respectively in 3-4 days, and 9-10 days after penetration. After the final moult, adult males and females can be encountered after 13 and 16 days respectively, after penetration of host. Feeding is necessary for the development of successive life stages. The number of generation that the nematode can complete within a year appears to be variable depending upon agro-climatic conditions. The nematode may complete one or even 2 to 3 generations in a year[8,9].

H. mucronata penetrates rice roots for feeding and oviposits inside the cortex[10]. The second-stage, from eggs present in soil, do

TABLE 1

Geographical distribution of some of the species of *Hirschmanniella*

Species	Country
H. oryzae	Indonesia, Japan, Bangla Desh, U.S.A., Sri Lanka, Malayasia, Madagascar, Thailand, Philippines, Sierra Leone, Nigeria, Venezuela, El Salvador, Taiwan, Korea, Ghana, Senegal, Gambia, Mauritania, Vietnam.
H. spinicaudata	Cameroun, Venezuela, Nigeria, Ivory Coast, Senegal, Gambia and Cuba.
H. gracilis	East Germany, Taiwan, India, Netherland, France.
H. mucronata	India, Thailand, Bangla Desh, Philippines.
H. thornei	Indonesia, Cuba.
H. imamuri	Japan, Korea, Vietnam.
H. belli, H. marina, H. zostericola, H. pomponiensis	U.S.A.
H. loofi	Netherland, Italy, Poland, Cuba.
H. anchoryzae	Canad.
H. kaveri	Indiaa.
H. miticausa	Solomon islands.

not seem to have the same capacity to penetrate the roots as those which are released inside the roots[11]. In contrast to these from roots, the cephalic sclerotization and stylet in the emerging larvae from eggs laid in soil are relatively weak. The egg laying inside the roots is considered to be an adaptation in the nematode for ensuring suitable feeding sites for the newly emerged second-stage larvae. The sex ratio in the common species, *H. oryzae*, *H. mucronata* and *H. spinicaudata* is almost equal.

TABLE 2

Host range of some common species of *Hirschmanniella*

Species	Hosts	References
H. oryzae	Rice, Cyperus, *Echino crusagalli*, *Fimbristylis* sp., *Leptohoria chinensis*, *Eclipta alba*, *Crogophora* sp., *Brachiaria ramosa*	12, 13
H. mucronata	Rice	
H. spinicaudata	Rice and some weeds belonging to Graminae and Cyperaceae	14
H. miticausa	*Colocasia eseulenta*	15
H. anchoryzae	Sword grass, wild rice	16
H. gracillis	Rice, *Alnus glactinosa*, *Salix cinerea*	
H. areolata	Lily, water chestnut	16
H. imamuri	Rice	
H. pomponiensis	Bulrush	17

Of the several environmental factors, that affect survival of the nematodes between host crops, perhaps, the two most important factors are soil moisture and presence of host. *H. oryzae* remains alive even upto ten weeks or more in dry soil free of host-roots[12]. The survival period, however, is extended to 10 months when the nematodes remain inside roots and under dried conditions. In flooded rice fields, they can survive upto seven months[13]. Under field conditions. the nematode may remain quiescent in dried up clay soils, during dry season after harvest of rice crop[3]. When the soils become moist or wet, the nematodes emerge from the roots into soil to seek fresh infection. During winter, the parasite survives as larvae or adults inside dry rice roots. In very wet fields, they hibernate in egg stage but, in Japan, this species rarely survives in wet soil but can do so in well drained fields[14]. Apparently *H. oryzae* is capable of withstanding extreme moisture stress, involving either dry or wet conditions.

H. imamuri and *H. spinicaudata* exhibit almost similar behaviour. The latter species survives better in roots than in soil. *H. imamuri* also survives in rice roots under adverse conditions and comes out from the roots into soil at the time of planting of rice crop, in the month of June, to initiate fresh infection[15].

In most of the Asian countries, rice is grown as rainfed crop and under constant flooding. *Hirschmanniella* species are the only nematodes, under these conditions, which not only survive but also are well adapted to cope with this kind of situation. Japanese populations of *H. oryzae* show extreme specialization, attacking the upland rice crops only when the transplants are subjected to flooded conditions[16,17]. This species also occurs in poorly drained rice soils under which condition most of the parasites fail to survive.

Informations are lacking regarding effect of soil types on the survival and multiplication of the nematodes. *H. oryzae* and *H. spinicaudata* appear to thrive better in clay and heavy clay soils[18]. Little is known about effect of temperature on the survival of these parasites. *H. oryzae*, *H. imamuri* and *H. spinicaudata* appear to thrive temperatures below 10°C. The optimum temperature for survival of the nematodes are 20-25°C which coincides with rice growing temperature in tropics[19]. *H. cryzae*, under laboratory conditions, when stored for 8 weeks at 15 and 35°C can survive to the extent of 57.6 and 90 per cent respectively, but not at −2°C[20]. The adults of *H. spinicaudata* are less vulnerable to extreme fluctuation of temperature than the second-stage larvae. All the species can withstand wide range of pH with best survival chances between pH 5 and 9[19].

The population development largely depends on the stage of host crop and environmental conditions. In general, the maximum populations, in roots, are observed at heading, irrespective of planting time[16]. In soil, population declines immediately after planting and increases after harvest when the rice roots start decaying[21]. Highest population of *H. mucronata* is observed between tillering and heading[22]. Recent studies indicate that population build-up depends also on intensity of rice grown within a given time[23]. Studies carried out in Vietnam have shown that root and soil samples from one and two rice crops exhibit twice more nematode numbers in two rice crops than one crop. Population densities in root system are least at post-transplanting and highest at heading time. Number of nematodes inside roots increases 20 22 times from transplanting to heading time in field, with both crop sequences.

Pathogenecity and Host-Parasite Relationships

Symptomatology

Although the role of *H. oryzae* in causing 'mentek' disease, in Indonesia, has remained rather controversial, the three common species, *H. oryzae*, *H. spinicaudata* and *H. imamuri*, have been

Fig. 1. Rice seedlings exhibiting stunting and chlorosis due to *Hirschmanniella oryzae.*

demonstrated experimentally to induce chlorosis within 60 days of nematode inoculations. The symptoms are rather non specific. Stunted growth, leaf chlorosis, reduced tillering, retarded plant growth delayed flowering and physiological disorder are some of the general symptoms in the disease-condition expressions exhibited by *Hirschmanniella*-infected plants (Fig. 1). The symptom expression depends on variety and age of crop as well as soil conditions and other environmental factors. Infection results in decreased tillering, reduced root and top weights, decay of primary roots and root lesions. In Japan, it is associated with 'Akiochi' disease of rice and is spectacular in soil with low pH[23]. Under this situation, the infected plants exhibit reduced tillering, loss in yield, retardation in growth and weight as well as browning of roots. The nematode interferes adversely with the oxidizing capacity of plants and induces colouration by iron oxide[24]. Reduction in total sugar, decrease in amino acids and liberation of phenols are some of the metabolic changes that have been recorded in *H. oryzae* infected rice crops[25]. Both *H. oryzae* and *H. spinicaudata* lower the rhizosphere acetylene reduction activity and are partially responsible for limiting non-symbiotic nitrogen fixation in submerged rice soils[26].

The loss in vigour and yield are usually more prominent in non-fertilized crops than the fertilized ones. In non-fertilized fields, the yield loss may be upto 31 per cent as against 19 per cent in fertilized plots[17]. Generally, the losses are estimated to range between 28 to 42 per cent[27].

H. mucronata has been recorded to decrease tillering as well as plant growth. In the United States of America, it causes decay of root tips of primary roots in soils with high pH values. By and large, the damage depends on the age of crop at the time of infection. One day old seedlings inoculated with 5,000 and 10,000 nematodes show respectively 50 and 70 per cent reduction in yield. However, yield loss is somewhat less (45 to 65 per cent) when 45 day old seedlings are inoculated with the same number of nematodes. Inoculations of 75 day old seedlings does not result in any yield loss[28]. Increase in sugars and amino acids are the characteristic metabolic changes in *H. mucronata* infected rice plants[25]. *H. spinicaudata* has been reported to be associated with yellowing of rice plants in Cameroun. Similar symptoms are known also in Upper Volta[6].

An undescribed species of *Hirschmanniella* is suspected to be associated with miti-miti disease of corms of *Colocasia esculenta* in Solomon Islands[29]. The first above ground symptom of this disease is wilting of older leaves which subsequently become chlorotic. The base of central leaves remains straight and the infected plants die prematurely. Infected corms are brown and necrotic and show red streak in cross section.

All the stages of nematode including males are capable of penetrating and feeding in roots. The general pattern of feeding consists of probing or exploration, stylet penetration, salivation and ingestion. Damage to root tip occurs in the course of invasion and feeding. These activities predispose the roots to colonisation by secondary micro-organisms as well[29]. The young larval stages as well as males can move in and out from the root more frequently than the gravid females. The nematodes prefer to feed around lateral root tips though they are capable to feed along entire root portion. *H. spinicaudata* and *H. imamuri* become coiled during salivation and ingestion and *H. oryzae* assumes an open '*c*' posture[29].

The nematodes are essentially cortical feeders, but occasionally some species may feed on vascular tissues[7,12,29]. *H. mucronata* and *H. oryzae* enter young roots through epidermis, at some distance distal to root tip and move through air channels in both distal and proximal directions. The depth of penetration is variable with species. After entering roots, the nematodes orient themselves parallel to long axis. Damage may be confined to 5 6 parenchymatous cells. In certain instances, one large round cell in the vicinity of feeding site has been observed. The thin lateral roots, with no air channel, are not affected. Occasionally, *H. orzae* may move near to the central cylinder, producing hollow cavities in the cortex during the process of feeding and movement.

The nematode can move intra- and inter-cellularly, causing extensive mechanical damage. Under axenic conditions, necrosis around invasion and feeding sites are not observed during the first five weeks after inoculation. The formation of large cavities, in cortical parenchyma of roof, is often accompanied by root invading micro-organisms which eventually leads to the death of the cell.

Control

Amongst the various control practices suggested, early planting of rice has been demonstrated in Japan to reduce the nematode populations. It has been shown also that the transplanted crops harbour large populations and suffer more damage than the directly seeded plants[30].

Reasonably good harvest can be ensured, even from heavily infested fields, if the crops are liberally fertilized. Application of calcium silicate, calcium nitrate, urea and compost keeps the soil population well below damage threshold level. Soil application of *neem* and mustard seed cake, at 220-240 kg/ha can improve plant growth and reduce the nematode population. Farm yard manure, water hyacinth, biogas effluents and *Ipomea* sp. are known to control soil population effectively. General sanitation, like removal of rice root stubbles after harvest, and keeping field free of weeds are some of the practices advocated for the control of the nematode. Rotation with peanut, jute, wheat, potato or black gram reduces the nematode population considerably[25].

Sources of resistance have been reported also from India and elsewhere. Cultivars R.P. 1155-128-1, CR 142-3-2, CR 52, N 136 and W 136 are recorded as exhibiting resistance against the nematode[25].

Chemicals like DD, DBCP, EDB, MBr, etc., have been demonstrated to be effective against these nematodes resulting in reduction in nematode population and increased yields upto 36 per cent. Granular chemicals like carbofuran (3% G), mocap (10% G), aldicarb (5% G), diazinon (5% G), phenamiphos (5% G), phorate (5% G), etc., have been demonstrated also as effective.

REFERENCES

1. Van Breda De Haan, J., Een aaltjee-ziekte der rijst 'omo-

mentek' of 'omo-bambang'. *Meded. Lds. Pltuin, Batavia,* 53, 65 pp., 1902.

2. Goodey, T., Soil and Fresh Water Nematodes. 2nd edition, revised by J.B. Goodey, London, Methuen, 544 pp., 1963.

3. Thorne, G., Principles of Nematology, New York, McGraw-Hill Book Co., 553 pp., 1961.

4. Van der Vecht, J., The problem of the mentek disease of rice in Java. *Pemb. Balai Besar Penj. Pert. Bogor*, 137 : 88 pp., 1953.

5. Ou, S.H., Rice diseases of obscure nature in Tropical Asia with special reference to 'Mentek' disease in Indonesia. *Int. Rice Common. Newsl.*, 14, 4-10, 1965.

6. Fortuner, R. and Merney, G., Root-parasitic nematodes of rice. *Revue Nematol.*, 2, 79-102, 1979.

7. Mathur, V.K. and Prasad, S.K., Embryonic development and morphology of larval stages of the rice-root nematode, *Hirschmanniella oryzae. Indian J. Nematol.*, 2, 146-157, 1972.

8. Kawahara, M. and Iyatoni, K., Studies on the bionomics of the rice root nematode, *Hirschmanniella imamuri* Sher and *Hirschmanniella oryzae* (Saltwedel) Luc & Goodey, with a special reference to its mode of life and population dynamics. *Jap. J. Appl. Ent. Zool.*, 14, 117-121, 1970.

9. Fortuner, R., Les nematodes parasites des racines associes an—*viz.*, an Senegal These Docteur—Ingenieur, Univ. Cl. Bernard, Lyon, France, No. 249, 51 pp., 1976.

10. Mahapatra, N.K., and Rao, Y.S., Effect of *Hirschmanniella mucronata* infestation on sprouting rice seed (Abstr.). *Proc. 62nd Ind. Sci. Cong.*, Delhi, Part III, 168 p. 1974.

11. Mahapatra, N.K. and Rao, Y.S., Feeding and digestion in the rice root-nematode, *Hirschmanniella mucronata. Proc. 62nd Ind. Sci. Cong.* Part III, 194 p., 1974.

12. Van der Vecht, J. and Bergman, B.H.H., Studies on the nematode *Radopholus oryzae* (Van Breda de Haan) Thorne and its influence on the growth of the rice plant. *Pemb. Balai Besar Penj. Perl. Bogor*, 131, 82 pp., 1952.

13. Mathur, V.K. and Prasad, S.K., Survival and host-range of the rice root-nematode, *Hirschmanniella oryzae*. *Indian J. Nematol.*, 3, 80-93, 1977.

14. Kawashema, K., Ecology of *Radoph lus oryzae* attacking rice roots. *J. Plant Prot.* Tokyo, 16, 57-59, 1962.

15. Gota, M., Invasion of rice roots in paddy field by the rice root nematode (*Hirschmanniella imamuri*) Abstract Comm. 17th Ann. meeting *Jap. Soc. Appl. Entomoi. Zool.*. Nagano, Japan 3-5 April, 1973, 128 p., 1973.

16. Fortuner, R., Etude ecologique des nematodes des rizieres du Senegal Cah. ORSTOM, *Ser Biol.* 11, 179-191, 1977.

17. Fortuner, R., Fertilisation du riz et degals ceuses par le nematode *Hirschmanniella oryzae* (Van Breda de Hann) Luc and Goodey, C.R. *hebd. Seans. Acad. Agri. Fr.* 58, 624-630, 1977.

18. Mathur, V.K. and Prasad, S.K., Occurrence and distribution of *Hirschmanniella oryzae* in the Indian Union with description of *H. mangaloriensis* n.sp. *Indian J. Nematol.*, 1, 220-226, 1971.

19. Babatola, J.O., Effect of pH, oxygen and temperature on the activity of *Hirschmanniella* spp. *Nematologica*, 27, 285-291, 1981.

20. Mathur, V.K. and Prasad, S.K., Survival and host range of the rice root nematode, *Hirschmanniella oryzae*. *Indian J. Nematol*, 3, 88-93, 1973.

21. Yokoo, T., Ookushi, K. and Su, W.C., On the change of nemic fauna and the population density in the paddy field before and after removing of the irrigated water especially the change of the population density of rice root of *Hirschmanniella oryzae*. *Agric. Bull. Sagu Univ.*, 23, 17-25, 1966.

22. Das, P.K. and Rao, Y.S., On the optimal sampling time for assessment of nematode population in rice soils. *Curr. Sci.*, 40, 17-18, 1971.

23. Kawashima, K., Studies on *Hirschmanniella oryzae* in the sub-reduction and the nematode injury. *Ann. Rebt. Soc. Pl. Prot.* North Japan, 15, 131-132, 1964.

24. Kawashima, K. and Fujima, T., On the injury to the rice plant caused by the rice root nematode (*Hirschmanniella oryzae*): Injury to the rice seedlings. *Res. Bull. Fukushima Pref. Agrl. Expt. Stn.*, 1, 57-64, 1965.

25. Rao, Y.S., Prasad, J.S. and Panwar, M.S., Nematode problems in rice: Crop losses, symptomatology and management. In Plant Parasitic Nematodes of India: Problems & Progress (Eds. Gopal Swarup and D.R. Dasgupta). IARI Publication, 497, pp. 1986.

26. Rinaudo, G. and Germani, G., Effect of the nematode *Hirschmanniella oryzae* and *H. spinicaudata* on N_2 fixation in the rice rhizosphere. *Rev. Nematol.*, 4, 171-172, 1981.

27. Taylor, A.L., Kaosiri, T., Sittaichi, T. and Baungsuwon, D., Experiments on the effect of nematodes on the growth and yield of rice in Thailand. *Pl. Prot. Bull.* F.A.O., 14, 17-23, 1966.

28. Panda, M. and Rao, Y.S., Evaluation of losses caused by the rice root nematode, *Hirschmanniella mucronata* Das, in rice (*Oryzae sativa* L.). *Indian J. Agric. Sci.*, 41, 611-614, 1971.

29. Babatola, J.O. and Bridge, J., Feeding behaviour and histopathology of *Hirschmanniella oryzae*, *H. imamuri* and *H. spinicaudata* on rice. *J. Nematol.*, 12, 48-53, 1980.

30. Nakazato, H., Kawashima, K. and Kurosawa, T., Seasonal occurrence of rice root-knot nematode. *Proc. Kanto-Tosan Pl. Prot. Soc.*, 11, 106, 1964.

14

Genus Pratylenchus Filipjev, 1936

The root lesion nematodes, *Pratylenchus* spp. constitute one of the major groups of migratory endoparasites, which occur throughout the world. Although these nematodes were recorded, as early as 1929, under generic names of *Tylenchus* or *Anguillulina*[1], the detailed description of a species, *P. pratensis*, was provided much later. The generic name, *Pratylenchus* was proposed by Filipjev[2] which was readily accepted by many workers. The economic importance of root lesion nematodes was recognized even during the early stage of nematology when Zimmermann[3] demonstrated pathogenicity of *T. coffeae* (=*P. coffeae*) on coffee plants by raising young seedlings on nematode-infested and clean soil. Since then, the role of several *Pratylenchus* species, as one of the limiting factors in crop production, has been established in a variety of plant species of economic importance, under different agro climatic conditions[4]. The first major taxonomic revision of the genus was made by Sher and Allen[5], and till to date 64 nominal species have been described from different parts of the world[6]. Of the described species, *P. penetrans*, *P. coffeae*, *P. zeae*, *P. thornei*, *P. vulnus* and *P. brachyurus* are considered to be of economic importance.

Morphology

These are relatively short and stout, measuring less than one mm long. Lip region is low, flattened, continuous with body contour and heavily sclerotized. Stylet measures 20 μm or less in length, with round, anteriorly flat or indented basal knobs. Oesophageal glands overlap intestine, mostly ventral. The median bulb of the oesophagus is massive and muscular. Vulva is posterior. Female has a mono-prodelphic ovary with only anterior ovary as functional. Post-vulval uterine sac is present which may or may not have rudiments of posterior ovary. The males are similar to females.

Distribution and Host Range

A number of species have a world-wide distribution. These include : *P. brachyurus*, *P. coffeae*, *P. crenatus*, *P. neglectus*, *P. penetrans*, *P. scribneri*, *P. thornei* and *P. zeae*[7,8]. *P. brachyurus*, *P. coffeae* and *P. zeae* are considered to be tropical and *P. penetrans*, *P. pratensis* and *P. neglectus* are judged to be temperate zone species. All species are polyphagous, having wide host range (Table 1) and capable of feeding and multiplying on variety of diverse plant species.

Biology

The life cycle is typical to that of many tylenchid plant parasitic nematodes with minor variations. Eggs are deposited singly into cortical cavities of infected roots. The maximum number of eggs laid by a single female of *P. penetrans* is about 68[40]. Embryonic development appears to be identical in all species. During the first moult (7 to 8 days after egg deposition), stylet and oesophagous become distinct. It takes about 10 days, from unsegmented egg to emergence of second-stage larvae in *P. penetrans* at 23°C[41]. All the larval-stages are known to feed. Feeding is considered to be necessary for development and deposition of eggs.

P. penetrans, *P. vulnus* and *P. coffeae* are bisexual species with amphimictic reproduction. The chromosome number in these three bisexual species are respectively 5, 6 and 7[42]. The monosexual *P. scribneri* is reported to have two chromosomal and reproductive forms. In the first form, n=6 chromosome and it reproduces with meiotic parthenogenesis. The second form has n=25 and reproduces by mitotic parthenogenesis. *P. zeae*, *P. brachyurus* and *P. neglectus* are also monosexual species with chromosome numbers as

TABLE 1

Host range and distribution of some *Pratylenchus* species

Crop	*Pratylenchus* sp.	References
(1)	(2)	(3)
Oat, Barley, Rye and Wheat	*P. crenatus, P. thornei, P. neglectus*	8, 9
Corn, Sorghum	*P. brachyurus,* P. zeae, P. penetrans,* P. hexincisus,* P. crenatus, P. thornei** and *P. neglectus**	9, 10-12
Cotton	*P. brachyurus*	13
Peanut	*P. brachyurus*, P. coffeae*	14-16
Potato	*P. penetrans, P. andenus, P. brachyurus,* P. coffeae, P. crenatus, P. neglectus, P. scribneri, P. alleni, P. thornei*	8, 17
Rice	*P. brachyurus, P. zeae*	8
Soybean	*P. alleni,* P. penetrans,* P. brachyurus,* P. scribneri, P. coffeae, P. crenatus, P. hexincisus,* P. zeae*	18-20
Sugarcane	*P. zeae, P. brachyurus, P. pratensis, P. delattrei, P. scribneri, P. coffeae, P. thornei*	21
Sweet potato	*P. brachyurus, P. zeae, P. pratensis, P. coffeae**	22, 23
Tobacco	*P. brachyurus,* P. hexincisus,* P. neglectus,* P. penetrans,* P. thornei,* P. vulnus,* P. zeae*,*	24
Carrot, Spinach,	*P. penetans,* P. pratensis,**	

(*Contd.*)

(Contd.)

(1)	(2)	(3)
Tomato, Limabean, Garden pea, Cabbage, Cowpea	*P. coffeae**	25, 26
Apple	*P. penetrans**, *P. vulnus* and *P. coffeae*	8, 27, 28
Cherry	*P. penetrans**, *P. pratensis*, *P. vulnus**	29, 30
Citrus	*P. coffeae**, *P. brachyurus**	30
Grapevine	*P. vulnus**	31
Peach	*P. vulnus*, *P. zeae*, *P. penetrans*, *P. brachyurus*	8, 32
Banana	*P. coffeae*, *P. goodeyi*, *P. penetrans*, *P. scribneri*, *P. thornei*	33-35
Coffee, Tea and Cocoa	*P. coffeae**, *P. brachyurus*, *P. loosi**	36-37
Pineapple	*P. brachyurus**, *P. penetrans**, *P. zeae**	38
Yam	*P. coffeae**	39

* Pathogenic species to the crops listed. Host range includes plant species that support nematode population and multiplication.

21-26, 30-32 and 20 respectively and reproduce by mitotic parthenogenesis. All the mitotic forms are considered to be polyploid[42].

The generation time for individual species depends on temperature and plant species involved (Table 2).

Environmental factors like temperature, moisture, soil texture, plant species and varieties and age of host govern multiplication of

TABLE 2

Generation time of some *Pratylenchus* species

Species	Host	Temperature (°C)	Duration of generation (Days)	References
P. penetrans	pea	28	35	42, 43
	cryptomeria	15	86	
		20	42-44	
		24	35	
		30	30-31	
	alfalfa	15	92	44
		25	37	
		30	30	
P. zeae	corn	5	no reproduction	11
		15	84	
		30	28-21	
P. brachyurus	corn	15	98	11

lesion nematodes. Interaction of the nematode species with one or more of the ecological factors influence the reproduction rate and survival of the nematodes. Generally, temperature and host are the most important factors which have profound influence on the build up of the nematode populations.

The thermal limits for reproduction and other activities for most of the species appear to be 5° and 35°C with an optima range of 20 to 30°C. In *P. coffeae*, which is essentially a tropical species, the optimum temperature for multiplication is reported to be 29.5°C on rough lemon, *Citrus jambhiri*[33]. A maximum number of 7,653 nematodes per g of root was observed under experimental conditions on sour lemon, after two months when initial inoculum consisted of 140 nematodes[33], The optimum temperature for invasion and multiplication of the same species on potato was observed between

25 to 28°C[45]. *P. penetrans*, a temperate zone species, multiplies maximum at 16°C on potato and at 24°C on corn[46]. P. *scribneri* and *P. alleni* reproduce best at 20°C but fail to do so at a temperature below 27.5°C on soybean cultivator Clark 63. This indicates a direct effect of host on reproduction of the lesion nematodes[43].

The temperatures that favour penetration of *P. penetrans*, *P. neglectus*, in highest numbers, in corn roots are respectively 20 and 30°C. However, *P. penetrans* is well adapted in northern cooler climate of Canada and North America. In alfalfa root, the females can penetrate roots at temperatures between 5 to 35°C and males and larvae between 10 to 30°C[47].

The degree of moisture stress also determines the activity and survival of root lesion nematodes. Cool and moist seasons favour activities of *P. penetrans* and the population development is greatest at moderate soil moisture tension (pF 2-3) than at low or high moisture tensions[48]. Similarly, *P. alleni* has been recorded more per unit length of soybean tap root in dry than wet regimes during the first week of colonization. Dry regime results in high male to female ratio[49].

The movement, penetration and reproduction of *Pratylenchus* species are most favourable in sandy to sandy loam soils in comparison to silty or heavy soils.

Symptomatology and Losses

The above ground symptoms of *Pratylenchus*-infected plants are non specific. Infected fields may show moderately to severely stunted plants in discrete areas with yellowish to chlorotic leaves (Fig. 1). Stunting of tops followed by gradual wilting and death, loss of primary roots, severe pruning of roots, reduced size of blossom, shrinking of grains and reduced plant growth are some of the symptoms associated with the infection of lesion nematodes (Fig. 2). The most characteristic symptom, however, is production of lesions on roots, the extent of which may vary with host species and varieties. Initially, the lesions appear as a tiny, elongate, water soaked spot which soon turn to brown to almost black. The lesions enlarge, gradually coalesce and ultimately may girdle the root giving an impression of constriction. Lesions may be large as

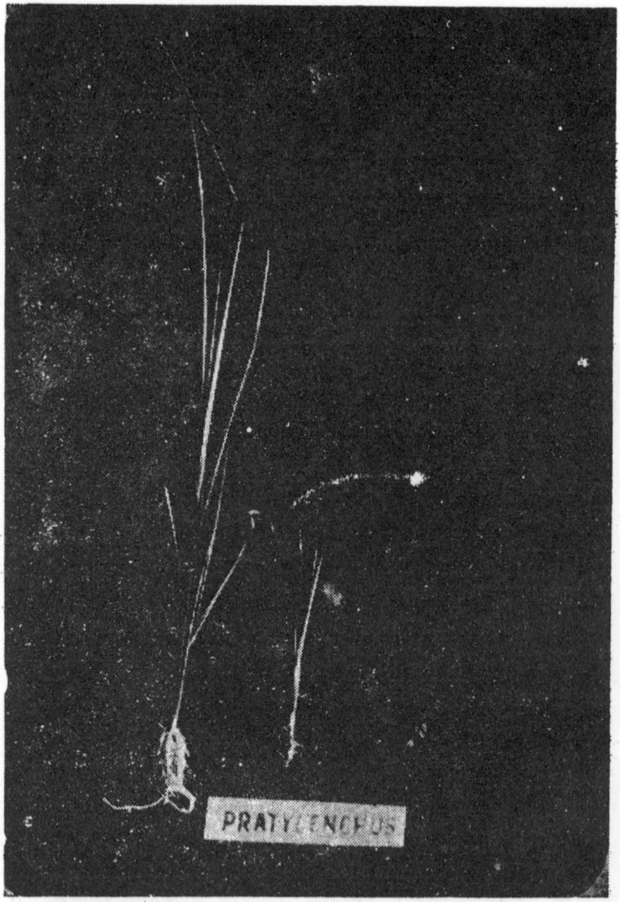

Fig. 2. Severe pruning of roots caused by *Pratylenchus indicus* on rice seedlings.

Left—Healthy seedling; Right—Infested seedling.

in peach infected with *P. penetrans*, which are formed within few hours of infection, or small which may be formed or observed several months after infection [27,50]. Fungi and bacteria may initiate secondary infection following nematode invasion and ultimately causing rot[51].

The damage may be mechanical or biochemical. All the species primarily cause cell necrosis and breakdown of root cortex. Mechanism of lesion formation has been explained recently on biochemical basis. The release of hydrolytic enzyme, glucosidase from *P. penetrans*, during extracorporeal digestion, results in hydrolysis of amagdylin into glucose, benzaldehyde and HCN. The toxic metabolites, benzaldehyde and HCN cause death of cell and necrosis.

In general, the necrotic symptoms are limited to feeding site and cavities in cortical region. The browning and necrotic symptom expression of infected host root is attributed due to accumulation of phenolic compounds in area of injury and also due to the activation of associated oxidative enzymes.

P. coffeae can be a serious threat to production of rough lemon seedlings or scioned stock. The adults as well larval stages of the nematodes are capable of infecting immature and mature tissues of *C. jambhiri*. The nematodes migrate inter and intra-cellularly within cortex in either direction from point of entry. Feeding and oviposition is restricted in cortex and cortical cavity respectively. In the case of entry of the nematode through the region of root tip, there is a complete destruction of meristem, eventually leading to suppression of root elongation. Occasionally, swelling with or without discolouration occurs in damaged root tip. Males, though feed and survive well within root cortex fail to produce any pockets. Multi-specific lesion nematode populations are not uncommon in citrus orchards. A mixed population consisting of *P. zeae*, *P. brachyurus*, *P. coffeae* and *P. penetrans* has been attributed to retard growth of citrus plants in Florida[35]. The feeding activity and movement of *P. brachyurus* within root cortex of soybean results in formation of large cavities and tunnels[30]. The damage to roots has been suggested due to mechanical injury made by intracellular movement of the nematode which results in rupture of the cell wall. Peanut is another crop which suffers loss in yield due to infection by *P. brachyurus*. The nematode feeds on root, pods and peg and adversely affects the yield and quality of the crop. *P. penetrans* causes extensive damage to cortical cell tissues and also injures endodermal cells of host roots. It is a pathogenic species on potato in Europe and North America. Recent field studies have demonstrated conclusively its role in causing growth suppression and consequently yield reduction (43-73 per cent) of potato[53]. The species is suspected also to be the primary agent accounting for decline for apple and other fruit trees.

Another important species, *P. vulnus* is widely distributed in grapevine soils of California (USA). The nematode causes lesions in the cortical parenchyma, girdles young rootlets and almost completely destroys fine root system[31].

Under field conditions, these nematodes, in quite a few cases, are associated also with micro-organisms which often increase the severity of disease expressions. However, in some cases, the nematode may provide the host protection from some fungal diseases. Such a situation has been recorded for black shank of tobacco where *P. penetrans* has been found to reduce the infection by the other organism[54]. Some of the recorded associations with other micro-organisms and the nature of disease development are listed in Table 3.

Control

Chemical

Aldicarb, phenamiphos, carbofuran, and methomyl have been tried, under field conditions, which have resulted in yield increases of various field crops, ornamentals and orchard plants. Preplant dips of propogating plant materials for 15 minutes in ethoprop (300-600 ppm), phenamiphos (200 ppm) and thionazin (150-300 ppm) reduce the population of *P. penetrans*[61]. Application of systemic chemical, oxamyl, as a foliar spray has been reported to give an excellent control of *P. penetrans* in the roots of apple seedlings[62].

Resistance

Use of plant resistance has been also promising in the management of these nematodes. The tobacco cultivar, PD 406, is resistant to *P. brachyurus*[63]. A number of soybean cultivars (Custer, Dyer, Forrest) show field resistance against *P. scribneri*[64]. Peanut shell tissues of Virginia Bunch 67 and GA 186-28 have been recorded to be less sensitive to *P. brachyurus* than the lines or cultivars of *Arachis hypogea*[65].

Fig. 1. Paddy field infested with *Pratylenchus indicus*.

TABLE 3

Synergistic effects of *Pratylenchus* species and other micro-organisms

P, ratylenchus sp. + micro-organism	Crop	Disease	References
P. penetrans + *Fusarium oxysporum* f. *pisi* Race 2	Pea	Enhanced wilting	55
P. vulnus + *F. oxysporum*	Peach	Additive effect on reduction of growth	56
P. penetrans + *F. oxysporum*	*Lotus corriculatus*	Reduction of foliage and root growth	57
P. brachyurus + *Rhizoctonia solani*	Soybean	Root damage	20
P. coffeae + *Acremonium stromaticum* and *Cylindrocarpon musae*	Plaintain	Root and rhizome rot	58
P. penetrans + *Aphanomyces euteiches*	Canning peas	Severity of damage	59
P. zeae + *Pythium graminicola*	Sugarcane	Severity of injury	60
P. penetrans + *Pseudomonas* spp.	Alfalfa	Growth reduction	8

Crop Rotation

All the species are polyphagous. Hence, choice of crops in rotational sequence becomes very difficult and problematic. Based on host efficiency tests, use of *Phalaris arundinaceae*, *Bromus inermis*, *Phleum pratense*, *Asparagus officinalis*, *Capsicum annuum* cv. California Wonder and Italian Sweet has been recommended wherever *P. penetrans* is the dominant population. Growing of these crops reduces the lesion nematode population[66].

Hot Water Treatment

Nematodes, residing inside dormant bulbs, corms, tubers, and fleshy roots can be controlled with hot water treatment. Strawberry plants infected with *P. penetrans* have been successfully treated with exposures in hot water at 49.4°C for 17.5 minutes or 52.8°C for 3 minutes or 54.4°C for 1 minute. In the case of flower bulbs, glasshouse crops and ornamentals, 43.3 and 46°C for 60 and 20 minutes respectively have been found to be quite satisfactory in significantly reducing nematode infestation.

Other Methods of Control

Growing of *Tagetes patula* cv. Harmony, in autumn after main crop or between rows of main crop has been suggested in the Netherlands[67]. The bacterim, *Pasteuria penetrans* has been demonstrated also to reduce soil population of *P. scribneri*[68].

REFERENCES

1. Goffart, H., Beobachtungen uber *Anguillulina pratensis*. *Zeitschr. Parasitenk.*, 2, 97-120, 1929.

2. Filipjev, I.N., The classification of free living nematodes and their relation to the parasitic nematodes. *Smithsonian Misc. Collec.* (3216) 89(6), 1-63, 1934.

3. Zimmermann, H., De nematoden der koffiewortels. *Meded pl. Tuin Batavia* 27, 16-41, 1898.

4. Thorne, G., Principles of Nematology. McGraw-Hill Publications, N.Y., Toronto, Lond., 553 pp. 1963.

5. Sher, S.A. and Allen, M.W., Revision of the genus *Pratylenchus* *Univ. Calif. Publ. Zool.*, 57, 441-570, 1953.

6. Siddiqi, M.R., Tylenchida : Parasites of plants and Insects. CAB Publication, U.K., 645 pp. 1986.

7. Corbett, D.C.M., *Pratylenchus pinguicaudatus* n.sp. with a key to the genus *Pratylenchus*. *Nematologica*, 15, 550-556, 1969.

8. Thames, W.H., The genus *Prutylenchus*. In Nematology in Southern Region of the United States. *So. Cop. Ser. Bull.*, 276, 1982.

9. Kort, J., Nematode diseases of cereals of temperate climates. In Economic Nematology (Ed. Webster, J.M.), Acad. Press, N.Y. & Lond., pp. 97-143, 1972.

10. Chevres-Roman, R., Gross, H.D. and Sasser, J.N., The influence of selected nematode species and number of consecutive plantings of corn and sorghum on forage production, chemical composition of plant and soil, and water use efficiency. *Nematropica*, 1, 40-41, 1971.

11. Olowe, T. and Corbett, D.C.M., Aspects of the biology of *Pratylenchus brachyurus* and *P. zeae*. *Nematologica*, 22, 202-211, 1976.

12. Zirakparvar, M.E., Host range of *Pratylenchus hexincisus* and its pathogenecity on corn, soybean and tomato. *Phytopathology*, 70, 749-753, 1980.

13. Endo, B.Y., Responses of root lesion nematodes, *Pratylenchus brachyurus* and *P. zeae* to various soil and plant types. *Phytopathology*, 49, 417-421, 1959.

14. Boswell, T.E., Pathogenecity of *Pratylenchus brachyurus* to spanish peanut. Ph.D. diss., Texas A & M Univ., 156 pp., 1968.

15. Miller, L.I. and Duke, G.B., Peanut nematode disease control. *Va. Agric. Exp. Stn. Bull.*, 520, 1961.

16. Good, J.M., Boyle, L.W. and Hammons, R.O., Studies on

Pratylenchus brachyurus on peanuts. *Phytopathology*, 48, 530-535, 1958.

17. Koen, H., Notes on the host range, ecology and population dynamics of *Pratylenchus brachyurus*. *Nematologica*, 13, 118-124, 1967.

18. Ferris, V.R. and Bernard, R.L., Injury to soybeans caused by *Pratylenchus alleni*. *Plant Dis. Reptr.*, 46, 181-184, 1962.

19. Taylor, P.L., Ferris, J.M. and Ferris, V.R., Quantitative evaluation of the effect of *Pratylenchus penetrans* on the root development of seedlings of soybeans under defined conditions. *Nematologica*, 14, 17, 1968.

20. Lindsey, D.W. and Cairns, E.J., Pathogenecity of lesion nematode, *Pratylenchus brachyurus* on six soybean cultivars. *J. nematol.*, 3, 220-226, 1971.

21. Williams, J.R., Nematodes attacking sugarcane. In Nematodes of Tropical Crops (Ed. Peachey, J.E.). *Commw. Inst. Helminthol.* St. Albans, Tech. Comm. 40, pp. 184-203, 1969.

22. Kikukawa, S. and Sakai, K., Studies on methods of breeding sweet potato varieties resistant to nematodes. *Helminthol. Abstr.*, 41, 4, 1969.

23. Birchfield, W., Hollis, J.P. and Martin, W.J., A list of nematodes associated with some Louisiana plants. *La. Agric. Expt. Stn. Tech. Bull.*, 101, 1978.

24. Lucas, G.B., Disease of tobacco. 2nd Ed., Scarecrow Press, N.Y., 165 pp., 1965.

25. Jensen, H.J., Experimental greenhouse host range studies of two root lesion nematodes, *Pratylenchus vulnus* and *P. penetrans*. *Plant Dis. Reptr.*, 37, 384-387, 1953.

26. Christie, J.R., Plant nematodes : Their bionomics and control. *Fla. Agric. Exp. Stn.* 256 pp., 1959.

27. Pitcher, R.S., Patrick, Z.A. and Mountain, W.B., Studies on the host-parasite relations of *Pratylenchus penetrans* (Cobb) to apple seedlings. *Nematologica*, 5, 309-314. 1960.

28. Mai, W.F., Parker, K.G., Controlling nematodes increases growth and yield of apples and cherries. *N.Y. State Hort. Soc. Proc.*, 115, 207-209, 1970.

29. McElroy, F.D., Nematodes of free fruits and small fruits. In Economic Nematology (Ed. Webster, J.M.), pp. 335-376, Acad. Press, N.Y. & Lond., 1972.

30. Parker, K.G. and Mai, W.F., Damage to tree fruits in New York by root lesion nematodes. *Plant Dis. Reptr.*, 40, 694-699, 1956.

31. Raski, D.J., Hart, W.H. and Kasimatis, A.N., Nematodes and their control in vineyards. Calif. Agric. Expt. Stn., Cir. 533, 1965.

32. Barker, K.R. and Clayton, C.R., Nematodes attacking cultivars of peach in North Carolina. *J. Nematol.*, 5, 265-271, 1973.

33. Radewald, J.D., O'Bannon, J.H. and Tomerlin, A.T., Temperature effect of reproduction and pathogenecity of *Pratylenchus coffeae* and *Pratylenchus brachyurus* and survival of *P. coffeae* in roots of *Citrus jambhiri*. *J. Nematol.*, 3, 409 416, 1971.

34. Brooks, T.L. and Perry, V.G., Pathogenicity of *Pratylenchus brachyurus* to citrus. *Plant Dis. Reptr.*, 51, 569-573, 1967.

35. Feldmesser, J. and Hannon, C.I., An evaluation of the role of *Pratylenchus* in the decline of citrus. *Phytopathology*, 55, 1058 (Abstr.), 1965.

36. Whitehead, A.G., Nematodes attacking coffee, tea and cocoa and their control. In Nematodes of Tropical Crops (Ed. Peachey, J.E.). *Commw. Inst. Helminth.*, St. Albans Tech. Comm., 40, 1969.

37. Sivapalan, P., Nematode pests of tea. In Economic Nematology (Ed. Webster, J.M.), pp. 285-311. Acad. Press, N.Y. & Lond., 1972.

38. Ayala, A.E., Gonzalez-Tejera, E. and Irizarry, H., Pineapple nematodes and their control. In Nematodes of Tropical Crops

(Ed. Peachey, J.E.), pp. 210-214. *Commw. Inst. Helminth.*, St. Albans Tech. Comm. 40, 1969.

39. Acosta, N. and Ayala, A.E., Pathogenecity of *Pratylenchus coffeae*, *Scutellonema bradys*, *Meloidogyne incognita* and *Rotylenchulus reniformis* on *Dioscorea rotundata*. *J. Nematol.*, 7, 1-6, 1975.

40. Mamiya, Y., Effect of temperature on the life cycle of *Pratylenchus penetrans* on *Cryptomeria* seedlings and observation on its reproduction. *Nematologica*, 17, 82-92, 1971.

41. Hung, C. and Jenkins, W.R., Oogenesis and embryology of two plant parasitic nematodes, *Pratylenchus penetrans* and *P. zeae*. *J. Nematol.*, 1, 352-356, 1969.

42. Roman, J. and Triantaphyllou, A.C., Gametogenesis and reproduction of seven species of *Pratylenchus*. *J. Nematol.*, 1, 357-362, 1969.

43. Dickerson, O.J., Pathogenecity, life cycle and morphology of certain *Pratylenchus* spp. on potato. Diss. Abstr., 1961.

44. Mai, W.F., Bloom, J.R. and Chen, E.D., Biology and ecology of the plant parasitic nematode, *Pratylenchus penetrans*. *Pa. Agric. Expt. Stn. Bull.*, 815, 1977.

45. Yokoo, T. and Kukoda, Y., On the relationships between invasion and the multiplication of the root lesion nematode (*Pratylenchus coffeae*) in the host plants and the soil temperature. *Agric. Bull.* Saga Univ., 22, 93-103, 1966.

46. Dickerson, O.J., Darling, H.M. and Griffin, G.D., Pathogenecity and population trend of *Pratylenchus penetrans* on potato and corn. *Phytopathology*, 54, 317-322, 1964.

47. Townshed, J.L., Economically important nematodes in Ontario. *Proc. Entom. Soc. Ont.*, 96, 15-16, 1966.

48. Kable, P.F. and Mail, W.F., Influence of soil moisture on *Pratylenchus penetrans*. *Nematologica*, 14, 101-122, 1968.

49. Norton, D.C. and Burn, N., Colonization and sex ratio of *Pratylenchus alleni* in soybean roots under two moisture regimes. *J. Nematol.*, 3, 265-269, 1971.

50. Mountain, W.B. and Patrick, Z.A., The peach replant in Ontario. VII. The pathogenecity of *Pratylenchus penetrans* (Cobb, 1917) Filip. & Stek., 1941. *Can. J. Bot.*, 37, 459-470, 1959.

51. Walia, R.K., The root lesion nematode *Pratylenchus*. In Plant Parasitic Nematodes of India : Problems and Progress (Eds. Swarup, G. and Dasgupta, D.R.), pp. 211-222, IARI Publication, 1986.

52. Acedo, J.R. and Rohde, R.A., Histochemical root pathology of *Brassica oleracea capitata* L. infected by *Pratylenchus penetrans* (Cobb) Filipjev and Schuurmans Stekhoven (Nematoda : Tylenchidae). *J. Nematol.*, 3, 62-68, 1971.

53. Olthof Th. H.A., Reaction of six *Solanum tuberosum* cultivars to *Pratylenchus penetrans*. *J. Nematol.*, 18, 54-58, 1986.

54. McIntyre, J.L. and Miller, P.M., Protection of tobacoo against *Phytophthora parasitica* var. *nicotianae* by cultivar nonpathogenic races, cell free sonicates and *Pratylenchus penetrans*. *Phytopathology*, 68, 235-239, 1978.

55. Seinhorst, J.W. and Kuniyasu, K., Interactions of *Pratylenchus penetrans* and *Fusarium oxysporum f. pisi* race 2 and of *Rotylenchus uniformis* and *F. oxysporum f. pisi* race 1 on peas. *Nematologica*, 17, 444-452, 1971.

56. Wehunt, E.J. and Weaver D.J., Effect of nematodes and *Fusarium oxysporium* on the growth of peach seedlings in the greenhouse. *J. Nematol.*, 4, 236 (Abstr.), 1972.

57. Willis, C.B. and Thompson, L.S., Influence of *Fusarium oxysporium* or *Pratylenchus* alone or in combination on growth and survival of birds foot trefoil. *Proc. Can. Phytopath. Soc.*, 37, 29 (Abstr.), 1971.

58. Pinochet, J. and Stover, R.H., Fungi associated with nematodo lesions in Honduras. *Nematropica*, 10, 112-115, 1980.

59. Oyekan, P.O. and Mitchel, J.E., The role of *Pratylenchus penetrans* in root rot complex of canning pea. *Phytopathology*, 62, 369-373, 1972.

60. Holtzman, O.V., and Santo, G.S., Effect of temperature on the interrelationship of *Pratylenchus zeae* and *Pythium graminicola* on sugarcane. *Phytopotholagy*, 61, 1321 (Abstr.), 1971.

61. Overman, A.J., Control of *Pratylenchus penetrans*, endoparasitic nematode of leather leaf fern. *Proc. Fla. Stat. Hort. Soc.*, 88, 573-578, 1975.

62. Abawi, G.S. and Mai, W.F., Effect of foliar application of oxamyl on movement of *Pratylenchus penetra s* in and outside the host roots. *Plant Dis. Reptr.*, 59, 795-799, 1975

63. Grahm, T.W., Tobacco varieties used to show differential response to root lesion and root-knot nematodes. *Plant Dis. Reptr.*, 49, 822-826, 1965.

64. Acosta, N., Malek, R.B. Edwards, D.I., Susceptibility of soybean cultivars to *Pratylenchus s ribueri*. *J. Agric. Univ. P.R.*, 63, 103-110, 1979.

65. Minton, N.A., Hammons, R.A., and Parham, H.A., Infection of shell and peg tissues of six peanut cultivars by *Pratylenchus brachyurus*. *Phytopathology*, 60, 472-474, 1970.

66. Miller, P.M., Reproduction, penetration and pathogenecity of penetrans on tobacco, vegetables and cover crops. *Phytopathology*, 68, 1502-1504, 1978.

67. Suatmadji, R.W., Studies on the effect of *Tagetes* species on plant parasitic nematodes. Diss. Agric. Univ., Wageningen, 1969.

68. Mankau, R., Biological control of nematode pest by natural enemies. *Ann. Rev. Phytopath.*, 18, 415-440, 1980.

15

Genus Radopholus (Cobb, 1893) Thorne, 1949

There are 22 recognised species under the genus with only one species, *R. similis*, as an economically important plant parasite. A sub-species, *R. similis citrophilus* Huettel, Dickson and Kaplan, 1984 is also of importance in some citrus orchards. The sub-species was earlier recognised as the 'citrus race' of *R. similis*.

Members of the genus exhibit sexual dimorphism in anterior region, with males having high rounded and offset cephalic region in comparison to females. Individuals range from 0.4—0.9 mm. in length. The males have indistinct and reduced oesophagus and spear.

The burrowing nematode, *Radopholus similis* occurs in most of the tropical and sub-tropical areas of the world. It has an extensive host range of 370 plant species causing considerable damage to many fruit, spice and agricultural plantation crops, in many tropical countries. A suggestion has been made about the possibility of co-evolution of the burrowing nematode alongwith black pepper or cultivars of banana that are known to have originated in the Western ghats of South India[1].

The nematode was first described in 1893, on the basis of a male specimen, obtained from necrotic roots of banana (*Musa sapientum*) growing in Fiji Islands[2]. It was named as *Tylenchus similis*. Some other names under which the nematode, recovered from other hosts, was described were : *Tylenchus granulosus* from banana on the basis of female specimens ; *Tylenchus acutocaudatus* from coffee roots in Java[3] and *Tylenchus biformis* from sugarcane in Hawaii[4]. Subsequently, Cobb[5] obtained specimens, collected around different host crops, and concluded that the nematodes observed first in Fiji, later in Hawaii and finally in Jamaica, were identical and same species. Accordingly. he retained *Tylenchus similis* as the type species and gave a detailed description of the nematode. The genus *Radopholus* was erected in 1949 to accommodate other similar species, retaining the type as *R. similis*[6]. A revision of the genus in 1969[7] suggests that the members of the genus appear to be indigenous to Australia and New Zealand since most of the species have beeen described from these countries[8].

Morphology and Biology

The burrowing nematode is a migratory endoparasite capable of spending its entire life in the roots. All larval stages and females infest healthy, succulent feeder root tips. Sexual dimorphism is apparent with degenerate, non-parastic adult males. The second-stage, male and female juveniles are differentiated by the presence of four specialised ventral chord nuclei, opposite the general primordium. This is exhibited in the females but not in males. In the third-stage juveniles, general increase in size and further differentiation of the gonad cells occur. At the fourth moult, males do not increase in length. The heavily sclerotised cephalic framework and strong stylet get replaced by an offset, unsclerotized head resulting in degenerated forms. Females emerge with fully formed gonads and the anterior gonad is usually longer than the posterior one. Eggs are laid singly inside the root and larvae hatch within 3-7 days. The life cycle takes 18-20 days at 24-26°C in citrus race and 20 25 days at 24-32°C in banana race[9].

The most important means by which *R. similis* is introduced into new geographical areas is through infested planting materials. In the present day fast trade between countries for foliage and other ornamental plants, such as *Anthurium* spp., *Calathia* spp., *Maranta* spp.

Pothos spp., etc., chances are more for introduction to newer areas through non-traditional hosts and also for introduction of new races into already known infested areas.

Soil and root samples for detection of *R. similis* are to be collected from the feeder root zone. Maximum population of *R. similis* is available on coconut, arecanut, black pepper and banana at the distances and depths shown against each crop within their root zones (Table 1).

TABLE 1

Distances and depths at which maximum numbers of *R. similes* are encountered

Crop	Distance from the base of the plant (cm)	Depth (m)
Coconut	100	50-100
Arecanut	25-75	25-75
Banana	25-50	20-40
Black pepper	25-50	20-30

Submersion of longitudinally split sliced roots in water, contained in petri dishes or shallow pans, at a temperature of 20-25°C, is ideal for extraction of *R. similis* from polyphenol rich roots of coconut, arecanut, banana and black pepper[1].

It is difficult to collect viable active nematode populations for culturing, etc., from black pepper roots by normal methods. Therefore, it is suggested to dissect individual lesions on roots in water, under a stereoscopic microscope and pick up nematodes to fresh water. The nematodes take a few minutes to show movements.

R. similis is easily cultured on okra roots[10], on root tissues developed from lucerne seed[11], on carrot discs[12] and on roots developed on leaf petioles of several citrus species and cultivars. It can be also cultured within the mesocarp of growing tender coconuts without affecting the size or quality of the nuts[13].

The host list of the citrus race of the burrowing nematode includes 246 species of citrus and non-citrus plants belonging to widely separated genera and families[14].

The citrus race is known to exist only in Florida and on *Anthurium* spp. in Hawaii[7]. Host range studies conducted in other geographical areas with banana race have added many hosts to the list making the total to 370.

Existence of two morphologically indistinguishable races of *R. simtlis*, namely, the 'banana race' parasitising banana and 'citrus race' parasitising both banana and citrus are known[15]. In addition to host preferences, citrus race has a haploid number of 5 chromosomes ($n=5$) against 4 in banana race. They also differ in their protein patterns and pheromone mediated behaviour (the males of banana race are attracted to females of citrus race, but do not copulate). These observations form the base to discard racial designation and elevation of the citrus race to a species rank, *R. citrophilus* with retention of the banana race as *R. similis*[16,17]. However, this has not received wide acceptance even now[18]. Further, there is a report of a *R. similis* population from Puerto Rico, with a haploid number of 5 chromosomes ($n=5$) and not infesting 5 species of citrus[19].

Highest populations of the nematode, in citrus groves on Lake land fine sand, occur from October to December and lowest from January to July. The population is known to disappear from known infested areas in spring[20,21]. In Central and South America, maximum population of banana, is recorded at inflorescence followed by a decline towards harvest. Factors favourable to nematode multiplication are mean soil temperature between 23 and 25°C and a water tension level close to zero[22]. In roots of coconut, arecanut and black pepper in Kerala (India), the population reaches its peak during September and the lowest during April to July[1]. Nematode populations, in individual palms/trees/vines, vary considerably during the low and high peaks depending upon the age, variety and condition of the plants infected.

The nematode survives and increases best in well drained sand and less in poorly drained ones. It can also withstand a wide range soil pH, but population densities are generally reduced in extremely acidic or alkaline soils.

There are evidences that the nematode can survive, on citrus as well as banana, up to a period of 17 months under conditions of clean fallow and deep ploughing. However, the population is reduced faster on exposure, of infested roots and rhizomes, to sunlight and moisture stress. Roots of felled coconut palms yield active nematode population, mostly females, up to a period of six months.

Symptoms

The burrowing nematode causes typical lesions and rotting of roots. It is known to cause many diseases and the most important ones are being dealt here.

Citrus species

Spreading decline of citrus is one of the important plant diseases caused by the burrowing nematode. Nematode infested trees become unprofitable with reduced vigour and less fruit production. In a 28 year old 'Marsh' grapefruit groves, the diseased trees gave an average yield of only four boxes/tree as against 17 from non-infested part of the grove[23]. The affected trees have fewer and smaller leaves with abundance of dead twigs and branches. The trees appear undernourished and wilt readily even during periods of mild moisture stress. Seasonal flushes of growth are weak with three to four leaves. Trees in the decline stage bloom profusely with sparse fruit set, resulting in low yields. The affected trees generally do not die, but live with decline symptoms for at least 15 years. However, they never resume normal growth and productivity[23,24].

Extensive feeder root damage takes place 50 cm below ground level in affected trees ; 20 to 30 per cent of feeder roots at depths of 25 to 75 cm and 90 per cent of feeder roots below 75 cm are destroyed[25]. Grape-fruit seedlings, on transplantation to infested soil, get badly stunted and chlorotic after nine months. The roots get reduced greatly in number and extent. Lesions of various sizes cover all the roots and root-lets. Young lesions are slightly discoloured, whereas old ones are brown and crumby. Some lesions coalesce to form large cankers with callused margins. Cankers have dark centres with firm brown margins[27].

Females and larvae enter growing feeder roots near their tips, in the region of cell elongation and root hair production. The nematodes feed on the cortical parenchyma cells and gradually burrow towards the stele, creating tunnels and cavities in the cortex and stele. They attack all types of parenchymatous tissues and enter the stele through the passage cells of endodermis. Large number of nematodes accumulate in the phloem-cambium ring region. This part of the root is often completely destroyed leaving a cavity filled with nematodes. Cell reactions involve hypertrophy and when the nematode penetrates pericycle, hyperplasia and tumour formation occur. Wound gum accumulates next to the stele in the invaded area, imparting tan to amber colour in the older portions of the lesions[27].

There is an increase in total free amino acids and amides in leaves and roots of infested grapefruit seedlings. However, these changes disappear on recovery of plants[10,26,27]. The phenolic content of roots and leaves of infested citrus trees is also affected. The amount of bound phenols is low in susceptible cultivars and high in tolerant ones[11].

On avocado (*Persica americana* Mill.), characteristic dark brown to black lesions, as small localised spots or extensive depressed lesions girdling the root for several inches in length, are formed with infestation of *R. similis*[12].

Banana (*Musa spacies*)

The burrowing nematode infestation in banana is of economic importance in Pacific Ocean Islands, Ivory Coast, Central and South America, Caribbean Islands, India and South-East Asia. The nematode invades banana root system and causes lesions and girdling. Heavy infestation results in stunted growth of plants, premature defoliation, unthriftiness and small bunches and fruits.

They topple over easily during wet and windy weather because of inadequate anchorage[13]. Such plants are susceptible to drought conditions. The economic life of plantations is reduced to one year only, requiring annual replantings[9,14,28].

Pre-adult and adult female nematodes puncture epidermal cells along the entire length of the root. The nematodes enter the root through wounds and occupy an intercellular position within the cortex, one to four cells beneath the epidermis. The nematodes feed on the cytoplasm of nearby cells causing the cell walls to rupture. The nucleus is either ingested or degenerated. Such cell destruction results in formation of cavities in the cortex. If necrosis accompanies cell breakdown, the nematode either migrate through the the tissue in advance of the necrosis or move out of root into the soil to reinfest the root at another site[14]. The strongly suberised endodermis seems to serve as an effective barrier to the invasion of the stele Occasionally *R. similis* attacks leaf bases near the soil line[29].

The nematode is also associated with some fungal diseases of banana, either aggravating the disease symptoms or providing avenues for entry of weakly pathogenic fungi. The incidence of panama wilt, caused by *Fusarium oxysporum* f. *cubense*, has been shown to increase in presence of *R. similis*[30].

Black Pepper (*Piper nigrum* L.)

Black pepper was introduced to Indonesia from Kerala, India[31]. There is every likelihood that the burrowing nematode also got introduced alongwith the plants to Bangka Islands. The slow wilt disease of black pepper, prevalent in Kerala and Karnataka States (India) is identical with the pepper yellows of Bangka Island, in the sequence of symptom development. Both diseases have been demonstrated to be caused by the burrowing nematode. The nematode populations in Indonesia and Kerala have a haploid number of four ($n=4$) chromosomes[1,10]. The first indication of the yellows disease is the appearance of occasional yellowed leaves on vines. The number of such leaves increases gradually and within a year large portions or even the entire foliage becomes involved (Fig. 1). The yellowing is followed by leaf and spike shedding, cessation of

Fig. 1. Drying of leaves of black pepper due to infestation by *Radopholus similis*.

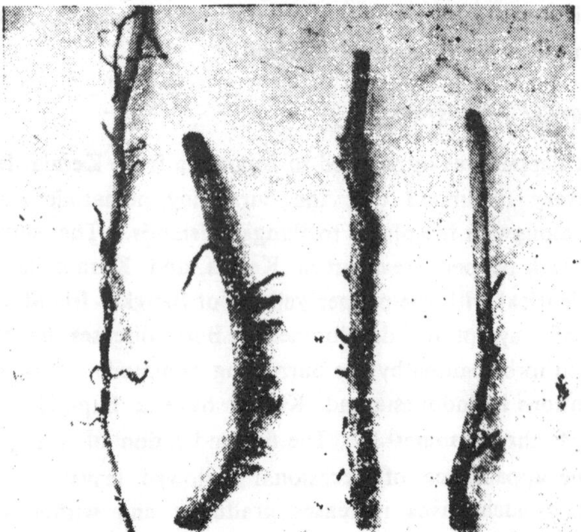

Fig. 2. *R. similis* infested roots devoid of feeder roots.

growth and die-back. Within two years of initiation of yellowing, all leaves are shed and death of the vine takes place. Young and old plants are affected and shedding of spikes is a major symptom of bearing vines. Large numbers of shed spikes are seen at the base of affected mature vines. The roots show numerous lesions and rotting. The main roots are devoid of fine feeder roots because of their quick rotting (Fig. 2). Extensive necrosis of larger lateral roots develop subsequently. The disease spreads fast to the adjacent vines. In a large plantation, affected patches become conspicuous with large number of barren standards or standards supporting hanging vines without any leaves or vines with varied intensities of yellowing of foliage. As high as 75-94 per cent reduction in plant growth has been reported after 150 days of inoculations with the nematode infestation load of 2,300 nematodes.

The nematode penetrates the roots within 24 hours of inoculation and the cells around the site of penetration become brown. Nematodes do not enter the steler portion[32]. Histological sections have shown presence of nematodes in inter and intracellular positions within the cortex. The xylem vessels are seen plugged with a gum-like substance[33]. The onset of yellows disease in Sumatra (Indonesia) is correlated with a nematode population of ?/100 g of soil or 25/10 g of roots of infested vines[34].

Ginger (*Zingiber officinale* L.)

Stunting, loss of vigour, with the top most leaves becoming chlorotic with scorched tips are the main symptoms of disease. Reduced tillering is apparent and affected plants tend to mature and 'dry down' faster than non-affected plants. The dormancy period of buds on infected rhizomes is short and they sprout early. Infections on rhizome are evidenced by small, shallow, sunken, water soaked lesions. These sites, on invasion by secondary organisms, lead to severe rotting of entire rhizomes. Histological examinations of infested rhizomes show presence of infective stages of *R. similis* even up to 1 cm in depth. Nematodes penetrate by passing through cell walls and are seen coiled within a single cell, having consumed the entire contents[35]. An initial inoculum levels of 10,000 nematodes per plant, over a period of six months, causes as high as 73 per cent reduction in rhizome weight[36].

Turmeric (*Curcuma longa* L.)

The symptoms are more or less similar to those obseıved in ginger. Drying of leaf tips and margins are conspicuous. Infested rhizomes are marked by the presence of shallow, water-soaked areas with rotting of roots. Scale leaves also harbour *R. similis.* An initial inoculum level of ten nematodes causes 35 per cent reduction of rhizome weight after four months and 46 per cent at the end of of the season (8 months)[37].

Coconut (*Cocos nucifera* L.)

Burrowing nematode-affected coconut palms exhibit general decline symptoms such as yellowing, reduction in size of leaves, button shedding, reduced yield, etc. These symptoms are not very specific and the only definite method to identify an infested palmis to look for symptoms on fresh white to orange coloured tender main roots.

It produces elongate, orange coloured lesions on tender roots. These lesions enlarge and coalesce to cause extensive rotting of roots. Tender roots on heavy infestation become spongy in texture. On merging of lesions, surface cracks develop on the semi-hard orange coloured roots (Figs. 3, 4). As high as 4000 nematodes can be recovered from one gram (one inch length) of main roots. Lesions and rotting are confined to the tender portions of roots and are not seen on the hard, older, red or brown coloured portions of the root. Lesions are also not conspicuous on the tertiary roots as they are very thin and rot quickly. An initial inoculum level of 62,500 nematodes per seedling causes respectively, 48, 22, 76, 18, 25, 40, 48 and 79 per cent reduction with respect to height, girth at collar region, shoot weight, number of leaves, number of leaflets/leaf, leaflet length, lamina length and root weight over control plants. Effect of parasitisation of the nematode is more pronounced on the root system, especially on the number and mass of feeder roots. The threshold inoculum density required for causing significant reduction

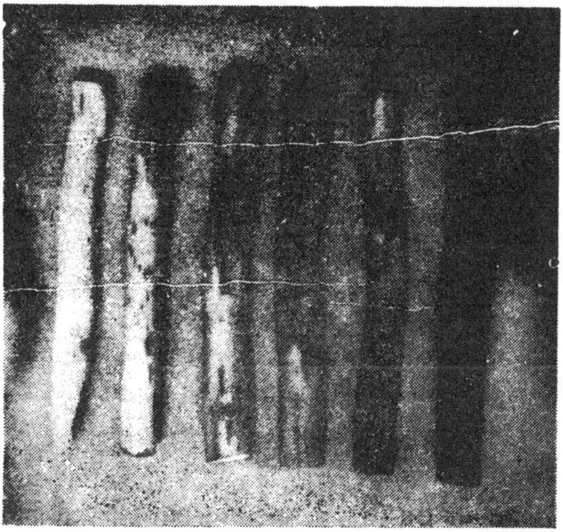

Fig. 3. Semi-hard orange coloured lesioned roots of coconut.

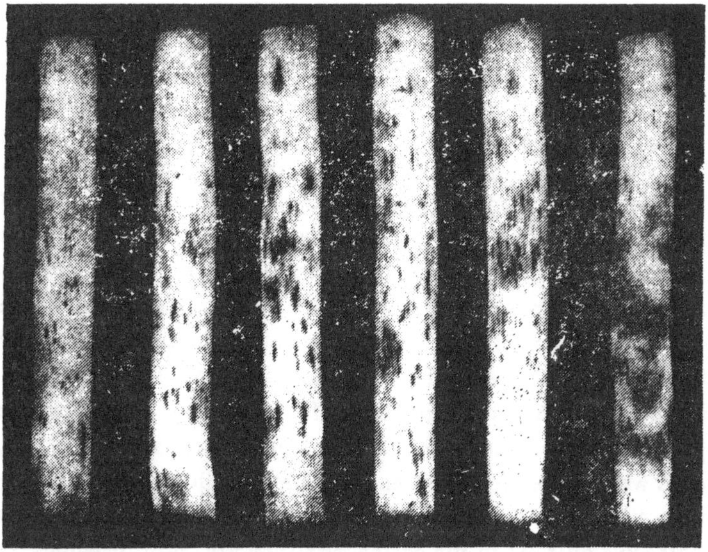

Fig. 4. Typical *R. similis* lesions on coconut roots.

of various growth parameters is 100 nematodes in 625 cm^3 or 900 g soil, under field conditions, over a period of five years.

Leaf bases and haustoria of seedlings are also infested by nematodes. Fungi, *Cylindroearpon effusum*, *C. lucidum* and *Cylindrocladium clavatum*, have been recorded in association with lesions, produced by *R. similis* on coconut roots. No appreciable damage is noticed in plants inoculated with the fungus, *Cylindrocarpon effusum*, alone. Plants inoculated simultaneously with the fungus and nematode, have recorded reduced rate of multiplication of the nematode.

Heavily infested seedlings, on field transplantation, fail to put out fresh leaves or new growth and fail to establish even under mild drought conditions.

R. similis infestation causes increase in total phenols, ortho-dihydroxy phenols, phenol oxidasing enzymes, total proteins and amino acids and decrease in total sugar, reducing sugar and starch contents in roots of coconut seedlings[1].

Arecanut (*Area catechu* L.)

The infested areca palms exhibit general yellowing and visible reduction in growth, vigour and yield. The nematodes produce small, elongate lesions on the young succulent, creamy white to light orange coloured portion of the main and lateral roots. Subsequently, the adjoining lesions coalesce and cause extensive root rot. The thick primary roots, produced from the bole region of the palm, exhibit large, oval, sunken dark lesions (nearly brown to black) from 2 mm to 2 cm in length. Unlike coconut, the tips of lateral and tertiary roots, on infestation, become black and are full of nematodes. An initial inoculum level of 100 nematodes per seedling or one nematode in 800 g of laterite soil causes visible damage ; and the percentage reductions over control are 23, 39, 25, 19 and 38 with respect to shoot length, shoot weight, girth at collar region, root length and root weight in three years. The nematodes cause extensive burrows in the cortex and they are seen oriented in inter and interacellular positions in transverse as well as longitudinal sections of the infested roots[38,39].

The fungus, *Cylindrocarpon obtusisporum*, is consistently found in association with lesions produced by the burrowing nematode on roots of areca palms. On concomitant inoculations, the fungus has been fund to inhibit multiplication of the nematode.

Coffee (*Coffea* **spp.**)

The nematode has been reported as a dangerous parasite of arabica coffee in Java[40,41]. The nematodes causes burrows in the cortex which turns yellow, then brown and finally rot. The roots sometimes get thickened and the bark becomes very rough.

Tea (*Camellia sinensis* **L.**)

The nematode was first reported as a pest of tea in Indonesia (Java)[8,42,43]. It causes premature flowering and fruiting together with a general stunted appearance of young tea plantings at altitudes of 500-100 m in Sri Lanka[44].

Groundnut (*Arachis hypogea* **L.**)

The nematode produces dark brown to black external cortical lesions, of irregular shape and size, on roots, underground stems, pegs and pods of groundnut except on the bacterial nodules. The nematodes infest the shell (exocarp and endocarp) and harbour the space between the endocarp and the testa of the kernel. Slightly more than 10,000 *R. similis* have been recovered from a fruit carrying severe shell lesions[45]. The banana race is more pathogenic to peanuts than the citrus race. Spanish and Valencia types are more susceptible than the Virginia type. The nematode do not survive the general curing and storage conditions of 24°C and 50 per cent RH for 7 days. Hence, the risk of seed transmissibility is negligible[46].

Sugarcane (*Saccharum officinarum* **L.**)

The nematode enters young sugarcane roots and feeds on the cortex. The root tissues lose their white or colourless appearance and take a cinnabar red colour, changing to reddish purple and finally turn to purplish black colour[5]. As high as 2,532 burrowing nematodes have been reported from one linear inch of sugarcane root in Hawaii[47]

Edible canna (*Canna edulis* **Ker.**)

The rhizomes of edible canna on infection produce irregularly shaped dark brown to nearly black spots, varying from a few millimeters to five cm in diameter and from two to ten millimeters in depth. The surface of the spot becomes hard, tough and leathery and gets flattened in contrast to the uniformly rounded surface of unaffected area. Roots easily break off in the affected region and sometimes rot completely[48].

Control

Control of the nematode is quite difficult because of its endoparasitic habit, its ability to survive on various host plant species and its concentration in deeper soil layers. However, in the case of citrus, the spreading decline disease is rather limited in distribution and it has been possible to devise uniform control measures[49, 50]. The same can not be said about the [populations identified with banana which are widely distributed all over the world. A number of chemicals listed in Tables 2 and 3, have been tested against the nematode population on various crops and which have proved to be quite effective, in some cases, under field conditions.

Hot Water Treatment

Being an endoparasite, 'hot water treatment' of the planting material is an excellent method to control the burrowing nematode. Various practices of heat therapy, adopted by different workers, with regard to banana sets and citrus seedlings, have been given in Table 4.

Flooding and fallowing have given good results in the banana fields[105]. Two years fallowing with a cover crop, *Panicum maximum* var. *trichoglume* is also a recommended practice in Australia[106]. In Kerala (India), 'sun drying' of rhizomes for 2-3 hrs, before planting, controls the nematodes to a great extent.

Varietal Resistance

Resistant sources have been exploited profitably in some cases. In *Citrus* species, 'Milam lemon', 'Ridge pineapple' sweet orange and 'Estes' rough lemon have been used as effective biological barriers[114–117]. In banana, tetraploids resulting from wild diploids

TABLE 2

Effect of various chemicals as 'Dip treatment' in control of *R. similis*

Chemical	Dosage	Duration and type of treatment	Effectiveness	References
(1)	(2)	(3)	(4)	(5)
(*a*) ***Banana***				
Carbofuran	750 ppm (75% wp)	10 minutes dip	Infection eliminated	51
	42 g a.i./sucker	(*i*) Immersion of suckers in clay slurry, uniform nematicides sprinkling	Significant reduction in nematode population and increase in bunch weight	52
		(*ii*) Paring followed by pralinage	Significant reduction in nematode population and increase in bunch weight	52

(*Contd.*)

(Contd.)

(1)	(2)	(3)	(4)	(5)
DBCP	Combined with Bordeaux mixture	Root dip	Reduced the infection from 89 to 1% after eight months	53
	0.05% emulsion	20 minutes dip after paring	Eliminated infection and promoted healthy root growth	54,55
	1% suspension	Five minutes dip	Eliminated infection	56
	550 ml+40 l clay soil+50 l water	Dip in the suspension for few seconds	No attack or spread or multiplication till 14 months	57
	0.5% emulsion	Five minutes dip (sets kept moist 24 h)	Eliminated infection	58
Ethoprophos	5000 ppm	One minute dip	Eliminated infection	59
	600 ppm	Twenty minutes dip	Eliminated infection	59
Fensulfothion	750 ppm	Ten minutes dip	Eliminated infection	5
	17 g a.i./sucker	Immersion of suckers in clay slurry, uniform nematicide sprinkling, paring followed by pralinage	Reduced population significantly and increased bunch weight	52

Oxamyl	1500 ppm	Ten minutes dip	Eliminated infection	51
	1000 ppm	20 minutes dip	Eliminated infection	59
Parathion	0.03% solution	Twenty minutes dip after peeling	Reduced the nematode population and increased the plant height	55
Phenamiphos	8 to 11 ml/plant 50 ppm (v/v) (40% emulsion)	Fifteen and thirty minutes dip	Eliminated infection	60
	100 ppm (40% emulsion)	Five minutes dip	Eliminated infection	58
	250 ppm (40% emulsion)	Ten minutes dip	Eliminated infection	59
	2500 ppm	Thirty minutes dip followed by a 30 mts. water rinse	Reduced infection from 71 to 14%	61
Phorate	8 to 11 ml/plant 50 ppm (v/v) (40% emulsion)	15 and 30 minutes dip	Eliminated infection	60
(*b*) *Citrus*				
Fensulfothion	1000 ppm	30 and 60 minutes dip	Eliminated infection	62,63

(*Contd.*)

(*Contd.*)

(1)	(2)	(3)	(4)	(5)
Phenamiphos	250 ppm	30 minutes dip	Eliminated infection	62
	400 ppm	30 and 60 minutes dip	Eliminated infection	63
	600 ppm	30 and 60 minutes dip	Eliminated infection	63
	1000 ppm	30 and 60 minutes dip	Eliminated infection	63
Prophos	1000 ppm	30 and 60 minutes dip	Eliminated infection	62
	1000 ppm	30 and 60 minutes dip	Eliminated infection	63
Thionazin	1000 ppm	30 and 60 minutes dip	Eliminated infection	62
	1000 ppm	30 and 60 minutes dip	Eliminated infection	63
(*c*) *Coconut*				
DBCP	1000 ppm	15 minutes dip	Eliminated infection	64
(*d*) *Ornamentals*				
Organic mercurial	1%	Dip for one hour	Eliminated infection	65

TABLE 3

Effect of various chemicals in the control of *R. similis*, under field conditions

Chemical	Dosage	Frequency of treatment	Effectiveness	References
(1)	(2)	(3)	(4)	(5)
(*a*) *Banana*				
Aldicarb	—	Three applications at 6 monthly intervals	Gave better control, decreased toppling	66
	60 g a.i./single mat	Twice a year	Reduced population significantly	67
	6 g a.i./mat	Four times a year	Reduced population significantly	68
	3 g a.i./mat	Three to four times a year	Reduced nematode population and increased yields	69

(*Contd.*)

(Contd.)

(1)	(2)	(3)	(4)	(5)
	4 g a.i./plant	At planting and 4 months later	Reduced population, produced earlier and improved yields	70
	30, 45 g/plant	Every 4 months	Controlled nematode, increased the yield and suckers	71
	30 g/plant	Every 6 months	Controlled nematode, increased the yield and suckers	71
	15 kg a.i./16 m^2	Alongwith weedicides	Controlled nematode	72
Aldicarb sulfone	50 g a.i./single mat	Twice a year	Reduced population significantly	67
	6.75 g a.i./mat	Three to four times a year	Reduced nematode population and increased yields	69
	9 g a.i./mat	Four times a year	Reduced nematode population and increased yields	68
	2.5 g a.i./mat	Three to four times a year	Reduced nematode population and increased yields	69

	4 g a.i. /plant	Thrice a year in a 35 cm band around the crown of each plant	Doubled the yield	73
	25 g a.i./hill	Every six months	Increased yield by 20-162%	74
Carbofuran	4 g a.i./plant	At planting and 4 months later	Reduced population, improved growth and yield	70
	6 g a.i./plant	At planting	Reduced population, increased plant growth under glass house conditions	75
	1 g a.i./plant	At planting and after 3 months	Reduced nematode population and increased yield	76
DBCP	40 l/ha	At planting during May-June, followed by 25 l/ha in October and 15 l/ha in March ; in each year thereafter	Controlled nematodes	77,73
	1·4 gal/acre	Injected at 10 points, one foot away from the base of the plant or at 15 points, 2 ft away	Reduced damage and increased fruit production	

(Contd.)

(Contd.)

(1)	(2)	(3)	(4)	(5)
		at 4 months' interval or through sprinkler irrigation		28
	—	Three applications at 6 months' interval	Gave better control, decreased toppling	66
	39.9 1/ha (75% EC)	Injected 15 to 25 cm deep at 8 points, 50 cm away from pseudostem initially and 10 months later to 18 weeks old plants	Produced heavier bunches	78
	3 ml/plant	Twice a year in 8 injections around the base of the plant, 20-25 cm away from the pseudo-stem	42% increase in crop growth and bunch weight	79
	3 g a.i./hill	Every six months	Reduced population significantly, 20-162% increase in yield	74

	300 1/ha	Before planting	Increased production by 30-50%	77
	2 g a.i./plant	At planting and 40 days later	Reduced population, increased plant growth under glasshouse conditions	75
EDB	150 kg/ha	Before planting	Increased production by 30-50%	77
Ethoprop	4-5 g a.i./mat	Three to four times a year	Reduced nematode population and increased yields	69
	4.5 g a.i./plant	Thrice a year in a 35 cm band around the crown of each plant	Doubled the yield	73
sazophos	2.5 g a.i./plant	Thrice a year in a 35 cm band around the crown of each plant	Doubled the yield	73
	2.5 to 3 g a.i./plant	Thrice a year	Controlled nematode	80
Methomyl	—	Three applications at 6 monthly intervals	Gave better control, decreased toppling	66
Oxamyl	3 g a.i./mat	Three to four times a year	Reduced population, improved plant growth and yield	69

(*Contd.*)

(Contd.)

(1)	(2)	(3)	(4)	(5)
Phenamiphos	2.5 g a.i./m^2	At 4 monthly intervals (April, August and December)	Reduced population significantly	81
	2 to 3 g a.i./plant	At 4 monthly intervals (April, August and December)	Reduced population significantly	82
	2 g a.i./plant	At planting and after 4 months	Reduced population significantly	83
	3 g a.i./mat	Three to four times a year	Reduced nematode population and improved growth and yield	69
	3 g [illegible]/plant	Thrice a year in a 35 cm band around the crown of each plant	Doubled the yield	73
	3 g a.i./hill	Every 4 or 6 months	Increased yield by 20-162%	74
	2, 4 or 6 g a.i./plant	Single application at planting	Reduced population,, improved plant growth under glasshouse conditions	75

Phorate	3 g a.i./sucker	A planting after paring	Eliminated infection under greenhouse conditions	84
(*b*) *Citrus*				
DBCP	4 U.S. gal./acre	Triple treatment in field, sprinkler application or injection or granules	Better control of infection	85
	10 U.S. gal./acre	Preplanting soil treatment in groves	Controlled infection for 4 years	86
	30 kg a.i./ha	Soil application thrice a year (March, May and April)	Controlled infection	87,88
DCP (Dichloropropene)	45, 60 and 70 U.S. gal./acre	Preplanting soil treatment in groves	Controlled infection for 4 years	86
EDB	83%—15 U.S. gal./acre	Preplanting soil treatment in groves	Controlled infection for 4 years	86,87
Ethion	750-1000 lbs/acre	Soil drenches on infected seedlings	Produced taller, heavier serial parts and heavier root system	90
Fensulfothion	100, 200 and 400 ppm	Sub-surface drench with water in nursery container	Eradicated nematode population in rough lemon seedlings	90

(*Contd.*)

(Contd.)

(1)	(2)	(3)	(4)	(5)
Metalaxyl	112 ppm	Preplantingtreatment under glasshouse conditions	Reduced nematode population, modified the roots to become tolerant	91
Oxamyl	11.4 kg a.i./ha	As foliar spray 6 times a year at 6 weeks intervals alongwith soil application of DBCP in March, July and May	Reduced nematode population	88
Phenamiphos	11.2 kg a.i./ha	Split application	Reduced population sigficantly	92
Prophos	100, 200, 400 ppm	Sub-surface drench mixed with water in nursery container	Eradicated nematodes in rough lemon seedlings	90
	8, 16 & 32 lbs a.i./acre	Three applications in greenhouse, in soil columns	Controlled nematode population upto 9' depth for 3 months	93

	60 lbs/acre, 16 lbs/acre	Three field applications	Increased twig growth	93
Thionazin	32 lbs/acre in 50 U.S. gal./water	Three treatments as soil drench around trees+post treatment water drench	Increased feeder roots	94
	100, 200, 400 ppm	Sub-surface drench mixed with water in nursery container	Eradicated nematode population in rough lemon seedlings	90
(c) Coconut				
(i) Nursery level				
Fensulfothion	50 kg a.i./ha	Thrice a year (in September, December and March)	Reduced burrowingnematode infestation	95
Phenamiphos	25 kg a.i./ha	Thrice a year (in September, December and May)	Eliminated infection in nursery seedlings	96
Phorate	25 kg a.i./ha	Thrice a year (in September, December and March)	Eliminated infection in nursery seedlings	96
(ii) Field Level				
Aldicarb	0.5-1.0 kg/ha	Applied once	Reduced nematode population, detected no resi-	

(Contd.)

(*Contd.*)

(1)	(2)	(3)	(4)	(5)
			dues in kernel and milk, 45 days after treatment	97
	10 & 50 g a.i./palm	Twice at 3 monthly intervals	Detected residues to the level of 0.07-0.38 ppm even 70 days afer second round	98
	10 g a.i./palm	One treatment	Detected residue level of 0.02 to 0.12 ppm 40-90 days after application	98
Phenamiphos	10 g a.i./palm	Twice a year in June-July and October-November	Increased yield by 30% and reduced nematode population	84
Phorate	10 g a.i./palm	Twice a year in June-July and October-November	Increased yield by 30% and reduced nematode population	84

(*d*) *Arecanut*				
(*i*) *Seedling stage*				
Aldicarb	1 g a.i./seedling	Thrice a year, for three years	Controlled soil and root population	99
Fensulfothion	1 g a.i./seedling	Thrice a year, for three years	Controlled soil and root population	99
(*ii*) *Adult palms*				
Aldicarb	10 g a.i./palm	Thrice a year (in June, September and January)	Reduced nematode population and increased yield	100
DBCP	10 ml a.i./palm	Thrice a year (in June, September and January)	Reduced nematode population and increased yield	100
Fensulfothion	50 g a.i./palm	Thrice a year (in June, September and January)	Reduced nematode population and increased yield	100
(*e*) *Black pepper*				
Aldicarb sulfone	4-8 kg a.i./ha	At nursery stage	Eliminated infection	101
DBCP		Field application	Effective in controlling slow wilt disease	102
Fensulfothion	4-8 kg a.i./ha	At nursery stage	Eliminated infection	101

(*Contd.*)

(Contd.)

(1)	(2)	(3)	(4)	(5)
Phorate	3 kg a.i./ha	Field application	Effective in controlling slow wilt disease	102
	3 g a.i./vine	Twice a year	Reduced nematode population and foliar yellowing	103
(*f*) *Ornamentals*				
(*i*) *Calathea*				
Aldicarb and Oxamyl	0.6 g/m^3	Monthly application around the plant	Eradicated nematode population in glasshouse	104

TABLE 4

Effect of hot water treatment on control of *Radopholus similis*

Temperature range	Duration of treatment	Type of treatment	Effectiveness	References
(1)	(2)	(3)	(4)	(5)
(*a*) *Banana*				
52-53°C	20 minutes	Immersion	Eliminated infection	107
53-55°C	20 minutes	Immersion	Eliminated infection	108
55°C	5 minutes	Immersion	Eliminated infection	109
55°C	15 minutes	Immersion+dip in organo-mercurial suspension	Eliminated infection	110
55°C	20 minutes	Immersion	Eliminated infection	25,56
55°C	25 minutes	Immersion	Eliminated infection	111

(Contd.)

(Contd.)

(1)	(2)	(3)	(4)	(5)
55°C	30 minutes	Immersion	Eliminated infection	112
60°C	10 minutes	Immersion+dip in organomercurial suspension	Eliminated infection	110
65°C	5 minutes	Immersion	Eliminated infection	113
(b) *Citrus*				
122°F (50°C)	10 minutes	Immersion followed by cooling for 10 minutes	Improved plant growth	65
122°F (50°C)	10 minutes	Frequent watering of the planting sites	Improved plant growth	49

and Gross Michel or Highgate triploids have shown considerable resistance to *R. similis* infestation[118]. In other crops, marked resistances have not been reported so far.

In-depth studies carried so far indicate that the nematode populations present on various crops, except citrus, belong to the banana race. The citrus race is confined to Florida (USA) only. However, it may be present in Hawaii also.

REFERENCES

1. Koshy, P.K., The burrowing nematode, *Radopholus similis* (Cobb, 1893) Thorne, 1949. In Plant Parasitic Nematodes of India : Problems and Progress (Eds. Swarup, G. and Dasgupta, D.R.), pp. 223-248, IARI Publication, New Delhi, 496 pp., 1986.

2. Koshy, P.K., Central Plantation Crops Research Institute, Krishna puram. In Nematology in India (Institutes and Universities), pp. 1-21, Indian Agricultural Research Institute, New Delhi, 163 pp., 1986.

3. Zimmermann, A., Het voorkomen van nematoden in de wortels van sirih en thee I. De aalt jesziekte der sirih (*Piper betle* L.) II. *Tylenchus acutocaudatus* Zn. in de thee, *Teysmannia*, 10, 523-531, 1899.

4. Cobb, N.A., Free-living nematodes inhabiting the soil about the roots of cane, and their relation to root diseases, (Fungus maladies of the sugarcane), *Hawaiian Sugar Planters' Assoc., Div. Path. and Physiol. Bull.*, 5, 163-195, 1906.

5. Cobb, N.A., *Tylenchus similis*, the cause of a root disease of sugarcane and banana. *J. Agric. Res.*, 4, 561-568, 1915.

6. Thorne, G., On the classification of the Tylenchida, new order (Nematoda: Phasmidia), *Proc. Helminth. Soc. Wash.*, 16, 37-73, 1949.

7. Sher, S.A., Revision of the genus *Radopholus* Thorne, 1949 (Nematoda : Tylenchoidea). *Proc. Helminth. Soc. Wash.*, 35, 219-237, 1968.

8. Colbran, R.C., Studies on plant and soil nematodes. Eleven new species of *Radopholus* Thorne and a new species of *Radhopholoides* de Guiran (Nematoda : Tylenchoidea) from Australia. *Qd. J. agric. and Anim. Sci.*, 27, 437-46), 1971.

9. Blake, C.D., Nematode parasites of banana and their control. In Nematodes of Tropical Crops (Ed. Peachey, J.E.). *Tech. Commun.* No. 40, *Commonw. Bur. Helminth.*, pp. 109-132, 1969.

10. Hanks, R.W. and Feldman, A.W., Quantitative changes in the free and protein amino acids in leaves of healthy and *Radopholus similis* infected and 'recovered' grapefruit seedlings. *Phytopathology*, 56, 261-264, 1966.

11. Feldman, A.W. and Hanks, R.W., Phenolic content in the roots and leaves of tolerant and susceptible citrus cultivars attacked by *Radopholus similis. Phytochemistry*, 7, 542, 1962.

12. DuCharme, E.P. and Suit, R.F., Nematodes associated with avocado roots in citrus spreading decline areas. *Plant Dis. Reptr.*, 37, 427-428, 1953.

13. Loos, C.A. and Loos, S.B., The black head disease of bananas (*Musa acuminata*). *Proc. Helminth. Soc. Wash.*, 27, 189-193, 1960.

14. Blake, C.D., Root rot of bananas caused by *Radopholus similis* (Cobb) Thorne and its control in New South Wales. *Nematologica*, 6, 295-310, 1961.

15. DuCharme, E.P. and Birchfield, W., Physiologic races of the burrowing nematode, *Phytopathology*, 46, 615-616, 1956.

16. Huettel, R.N., Dickson, D.W. and Kaplan, D.T., *Radopholus citrophilus* sp. n. (Nematoda), a sibling species of *Radopholus similis. Proc. Helminthol. Soc. Wash.*, 51, 32-35, 1984.

17. Huettel, R.N., Dickson, D.W. and Kaplan, D.T., Chromosome number of populations of *Radopholus similis* from North, Central and South America, Hawaii and Indonesia. *Revue Nematol.*, 7, 113-116, 1984.

18. Siddiqi, M.R., Tylenchida : **Parasites of Plants and Insects.** St. Albans, U.K., C.A.B., 646, 1986.

19. Rivas, X. and Roman, J., Investigations on the host-range of a population of *Radopholus similis* from Puerto Rico, *Nematropica*, 15, 165-170, 1985.

20. DuCharme, E.P. and Price, W.C., Dynamics of multiplication of *Radopholus similis*. *Nematologica*, 12, 113-121, 1966.

21. DuCharme, E.P. and Suit, R.F., Population fluctuations of burrowing nematodes in Florida citrus groves. *Proc. Fla. St. hort. Soc.*, 80, 63-67, 1967.

22. Hugon, R., Ganry, J. and Berthe, G., Population dynamics of the nematode, *Radopholus similis* in relation to the development of the banana and to the climate. *Fruits*, 39, 251-253, 1984.

23. Poucher, C.H., Ford, W., Suit, R.F. and DuCharme, E.P., Burrowing nematode in citrus. *Fla. Dep. Agric. Div. Pl. Ind. Bull.* No. 7, 63, 1967.

24. Suit, R.F. and DuCharme, E.P., The burrowing nematode and other parasitic nematodes in relation to spreading decline of cirtus. *Plant Dis. Reptr.*, 37, 379-383, 1953.

25. Ford, H.W., Effect of spreading decline disease on the distribution of feeder roots of orange and grapefruit trees on rough lemon rootstock. *Proc. Am. Soc., Hort. Sci.*, 61, 68-72, 1953.

26. Feldman, A.W. and Hanks, R.W., Quantitative changes in the free and protein amino acids in the roots of healthy, *Radopholus similis* infested and 'recovered' grapefruit seedlings. *Phytopathology*, 54, 1210-1215, 1964.

27. Hanks, R.W. and Feldman, A.W., Comparison of free amino acids and amides in roots of healthy and *Radopholus similis* infected grapefruit seedlings. *Phytopathology*, 53, 419-422, 1963.

28. Blake, C.D., The histological changes in banana roots caused by *Radopholus similis* and *Helicotylenchus multicinctus*. *Nematologica*, 12, 129-137, 1966.

29. Wehunt, E.J. and Edwards, D.I., *Radopholus similis* and other nematode species on banana. In Tropical Nematology (Eds. Smart Jr., G.C. and Perry, V.G.). *Univ. Fla. Press*, Gainesville, 1-19, 1968.

30. Newhall, A.G., The incidence of Panama disease of banana in the presence of the root-knot and burrowing nematodes (*Meloidogyne* and *Radopholus*). *Plant Dis. Reptr.*, 42, 853-856 1958.

31. Nambiar, M.C., Trends in Pepper cultivation. *Proc. National Seminar on Pepper*, Calicut, Dec. 19, 4-6, 1977.

32. Venkitesan, T.S. and Setty, K.G.H., Pathogenecity of *Radopholus similis* to black pepper (*Piper nigrum*). *Indian J. Nematol.*, 7, 17-26, 1977.

33. Freire, F.C.O. and Bridge, J., Histopathology of black pepper roots infected with *Radopholus similis*. *Fitopatologia Brasileira*, 10, 475-481, 1985.

34. Mustika, I., Observations on the relationship between nematode population and yellows disease on black pepper in Bangka. *Pemberitaan Lambaga Penelitian Tanaman Industri.*, No. 30. 11-12, 1978.

35. Vilsoni, F., McClure, M.A. and Butler. L.D., Occurence, host range and histopathology of *Radopholus simils* in ginger (*Zingiber officinale*). *Pl nt Dis. Reptr.*, 60, 417-420, 1976.

36. Sundararaju, P., Sosamma, V.K. and Koshy, P.K., Pathogenicity of *Radopholus similis* on ginger. *Indian J. Nematol.*, 9, 91-94, 1979.

37. Sosamma, V.R., Sundararaju, P. and Koshy, P.K., Effect of *Radopholus similis* on turmeric. *Indian J. Nematol.*, 9. 27-31, 1979.

38. Koshy, P.K. and Sundararaju, P., Pathogenicity of *Radopholus similis* on *Areca catechu* L. *Indian Phytopath.*, 36, 469-472, 1983.

39. Sundararaju, P. Studies on the burrowing nematode of arecanut. Ph.D. Thesis, Kerala University, Trivandrum, 133 pp. 1984.

40. Bally, W. and Reydon, G.A., De tegenwoordige stand van bet vraagstuck van de wortelaaljes in de koffie cultuur. *Arch. Koffiecult. Nederl.-Indie.*, 5, 23-216, 1931.

41. Zimmermann, A., De nematodes der koffiewortels. *Meded. pl. Tuin. Batavia.*, 27, 16-41, 1898.

42. Menzel, R., De plagen van thee in Nederlandsch-Indie (Java en Sumatra) en hare bestrijding. *Arch. Thee Cultuur*, 3, 11-64, 1929

43. Steiner, G. and Buhrer, E.M., The nematode *Tylenchus similis* Cobb, as a parasite of the tea plant (*Thea sinensis* L.), its sexual dimorphism and its nemic associates in the same host. *Z. Parasitkde.*, 5, 412-420, 1933.

44. Sivapalan, P., Association of *Radopholus similis* with decline in young tea fields. *Plant Dis. Reptr.*, 52, 528, 1968.

45. Martin, G.C., James, G.L., Bissett, J.L. and Way, J.I., Trials with field crops and *Radopholus similis* with observations on nematodes. *Rhod., J. agric. Res*, 7, 149-157, 1969.

46. O' Bannon, J.H., The influence of two races of the burrowing nematode *Radopholus similis* on peanut (*Arachis hypogea*), Nematology Circular, Division of Plant Industry, *Fla, Dept. Agric. Cons. Service* No. 112, 2 pp. 1984.

47. Cassidy. G., Nematodes associated with sugarcane in Hawaii. *Hawaii Plrs'. Rec.*, 34, 379-387. 1930.

48. Godfrey, G.H., The host plants of the 'burrowing nematode, *Tylenchus similis*' *Phytopathology*, 21, 315-322, 1931.

49. Cohn, E., Nematode diseases of citrus. In Economic Nematology (Ed. Webster, J.M.) London, Academic Press, 215-244, 1972.

50. O' Bannon, J.H. and Esser, R.P., Citrus decline caused by nematodes in Florida. IV. Management decisions. Nematology Circular, Division of Plant Industry. *Fla. Dept, Agric. Cons. Service* No. 118, 4 pp., 1985.

51. Canderon, O.P., Evaluation of four systemic namaticides in the treatment of banana rhizomes. *Fitopatologia*, 9, 41-42, 1974.

52. Rajagopalan, P., Chinnarajan, A.N. and Vadivelu, S., Control of banana nematodes by rhizome treatment. *S. Indian Hort.*, 25, 133-137, 1977.

53. Loos, C.A. and Loos, S.B., Preparing nematode free banana seed. *Phytopathology*, 50, 383-386, 1960.

54. Edmunds. J.E., Plant nematode problems of the Windward Islands. In Nematodes of Tropical Crops (Ed. Peachey, J.E.). *Tech. Common. No. 40, Commonw. Bur. Helminth.*, pp. 142-148, 1969.

55. Peachey, J.E. and Hooper, D.J., Chemical treatment of quarantined banana stocks infested with plant parasitic nematodes. *Pl. Pathol.*, 12, 117-120, 1963.

56. Casamayor, G.R., Seidel, D. and Decker, H., Tratamiento con agua caliente contra nematodes parasitos en platano. *Boln. Cent. Invesrnes Agropec. Univ. Cent. Las Villas*, No. 1, 6, 1966.

57. Vilardebo, A. and Robin, J., Nematicidal treatment of banana planting material. In Nematodes of Tropical Crops (Ed. Peachey, J.E.). *Tech. Commun. No. 40. Commonw. Bur. Helminth.*, 133-141, 1969.

58. Decker, H., Casamayor, G.R. and Rodriguez, F.M.E., Further investigations of ways and means on controlling parasitic nematodes in banana plants. *Beitrage zur Tropischen und Subtropis-*

chen Land Wirtschaft und Tropenveterinarmedizin, 9, 195-202, 1971.

59. Broadley, R.A., Nematicide treatment of banana planting material. *Aust. J. expl. Agric. Anim. Husb.*, 19, 631-633, 1979.

60. Taboada, J. and Caballero, D.J., Research on the chemical control of *Radopholus similis* (Cobb, 1893) Thorne, 1949 on banana plants. *An. Inst. Biol. Univ. Mex. Ser. Zool.*, 39, 29-33, 1968.

61. Jones, R.K. and Reynolds, R.E., The control of *Radopholus similis*, burrowing nematode, in banana planting material. *Subtropica*, 1, 9-12, 1980.

62. O' Bannon, J.H. and Taylor, A.L., Control of nematodes on citrus seedlings by chemical bare root dip. *Plant Dis. Reptr.*, 51, 995-998, 1967.

63. O' Bannon, J.H. and Tomerlin, A.T., Control of nematodes on citrus seedlings by chemical dips. *Plant Dis. Reptr.*, 55, 154-157, 1971.

64. Koshy, P.K. and Sosamma, V.K., Control of the burrowing nematode, *Radopholus similis* on coconut seedlings with DBCP, *Inaian J. Nematol.*, 9, 32-33, 1979.

65. Spears, J.F., Progress report of burrowing nematode control programme for calender year 1956, Hickville, New York, *USDA Agric. Res. Service*, Pl. Pest control, Br. IV, 38, 1955.

66. Coates, P.L., Response of an established banana planting to four nematicides. *Pest Arti. and News Summs.*, 18, 165-170, 1972.

67. Hasing-Lama, M., Franco, M.K. and Escobar-Goyes, Effects of aldicarb, UC 21865 and DBCP on *Radopholus similis* (Cobb) Infestations and yields in a banana plantation in Machala, Ecuador, *Acta Horticulturae*, No. 57, 213-221, 1976.

68. Figueroa, A., Effects of the nematicides, aldicarb, aldoxycarb and DBCP on nematode control and yield of bananas. *Nematropica*, 10, 15-20, 1980,

69. Boncato, A.A. and Davide, R.G., *Radopholus similis* on Cavendish banana in Davao del Norte. III Laboratory and field evaluation of nematicides for *R. similis* control. *Philipp. Agric.*, 63, 147-155, 1980.

70. Zem, A.C., Rodriguez, J.A.S., Alves. E.J. and Lordello, L.G.E., Effect of nematicides on nematode populations and productivity of banana var. Nanicao. Publication No. 5, 1981. *Sociedade Brasileira de Nematologia*, Piracicaba, SP, Brazil, pp. 57-70, 1982.

71. Roman, J., Oramas, D., Green, J. and Torres. A., Control of nematodes and black weevils in plantains. *J. Agric. Univ. P. Rico*, 67, 270-277, 1983.

72. Zem, A.C. and Alves, E.J., Effect of different cultural practices on the population of *Radopholus similis*. Publication No. 7, 1983, *Sociedae Brasiliera de Nematologia*, Piracicaba, SP, Brazil, 215-215, 1983.

73. Vilardebo, A., The application of research results for control banana nematosis caused by *Radopholus similis*. Cobb in West Africa. *Nematropica*, 11, 193-206, 1981.

74. Perez, J.A. and Mola, G., Use of Nemacur 10% G, Nemagon 20% G and Carbofuran 5% G to control *Radopholus similis* in Banana. *Cienca Y Tecnicia en la Agricultura, Proteccion de Planas*, Suppl., 7-18, 1982.

75. Ravichandra, N.G. and Krishnappa, K., Effect of various pesticides on the control of the burrowing nematode, *Radopholus similis* infesting banana. *Indian J. Nematol.*, 15, 26-29, 1985.

76. Charles, J.S.K., Venkitesan, T.S. and Thomas, Y., Comparative efficacy of antagonistic intercrops with carbofuran in control of

burrowing nematode, *Radopholus similis* in the banana cultivar Nendran, *Indian J. Nematol.*, 15, 241-242, 1985.

77. Luc, M. and Vilardebo, A.. Les nematodes associes aux bananiers cultives dan L'Ouest African. I. Esp'eces parasites dommages causes. *Fruits*, 16, 205-219, 1961.

78. Hutton, D.G. and Chung, D.C., Effects of post planting applications of the nematicide DBCP to Plantain. *Nematropica*, 3, 46-50, 1973.

79. Beugnon, M. and Vilardebo, A., The nematodes of the banana tree in Madagascar, aspects of the problem and its economic importance. *Fruits*, 28, 607-612, 1973.

80. Speich, J. Isazophos, a nematicide for the control of *Radopholus similis* in bananas. *Med. Fac. Landbouww. Rijksuniv. Gent.*, 47, 715-719, 1982.

81. Burnett, W.M., Burnett, R.M. and Inglis, I.M., The development of nemacur for control of nematodes in Australian banana plantations. *Pflanzesschutz Nachrichten*, Bayer, 27, 233-266, 1974.

82. Gowen, S.R., Effect of different nematicides on banana yields in the Windward Islands. *Nematropica*, 8, 9-13, 1978.

83. Nair, K.K.R., Studies on the chemical control of banana nematodes. *Agric. Res. J. Kerala*, 17, 232-235, 1979.

84. Koshy, P.K., Research on plant parasitic nematodes at the Central Plantation Crops Research Institute, Kasaragod, Kerala, 45, 1986.

85. Suit, R.F., DuChrame, E.P. and Feldman, A.W., Effectiveness of DBCP and fungicides for the control of *Radopholus similis* on citrus trees. *Plant Dis. Rep.*, 45, 62-66, 1961.

86. Suit, R F., A comparison of preplant soil treatment chemicals for control of burrowing nematode in citrus groves. *Plant Dis. Reptr.*, 45, 454-456, 1961.

87. Collins, R.J. and Feldman, A.W., Studies on the penetration

of DD, EDB and DBCP for the control of *Radopholus similis* and for the destruction of live citrus roots. *Proc. Fla. St. Hort. Soc..* 78, 63-69, 1966.

88. O'Bannon, J.H. and Tomerlin, A.T., Control of the burrowing nematode, *Radopholus similis* with DBCP and oxamyl. *Plant Dis. Reptr.*, 61, 450-454, 1977.

89. Tarjan, A.C. and Wouts, A.M., Stimulative and nematicidal effects of ethion on citrus seedlings parasitized by the burrowing nematode. *Proc. Fla St. Hort. Soc.*, 77, 60-66, 1965.

90. Taylor, A.L. and O'Bannon, J.H., Experiment with an apparatus for subsurface application of nematicides to nursery plants in containers. *Plant Dis. Reptr.*, 52, 218-222, 1968.

91. Kaplan, D.T., Influence of metalaxyl on three nematodes of citrus. *J. Nematol.*, 15, 454-460, 1983.

92. O'Bannon, J.H. and Tarjan, A.C., Management of *Radopholus similis* infecting citrus with DBCP or Phenamiphos. *Plant Dis. Reptr.*, 63, 456-460, 1979.

93. Suit, R.F. and Feldman, A.W., Treatment of citrus trees with cynem for control of *Radopholus similis. Plant Dis. Reptr.*, 45, 782-786, 1961.

94. Collins, R.J. and Feldman, A.W., Attempt to improve the growth of *Radopholus similis* infected citrus with under tree drenches of Zinophos. *Plant Dis. Reptr.*, 49, 865-867, 1965.

95. Koshy, P.K. and Nair, C.P.R., Control of *Radopholus similis* (Cobb, 1893) Thorne, 1949 in coconut nursery. *Indian J. Nematol.*, 9, 15-19, 1979.

96. Koshy, P.K., Sundararaju, P. Sosamma, V.K, and Ravikumar, P. Efficacy of four systemic nematicides against *Radopholus simils* in coconut nursery. *Indian J. Nematol.*, 15, 148-151, 1985.

97. Singh, K.P., Pandey, S Y., Khan, M.M. and Satendra Singh., Residues of aldicarb in coconut. *Pesticide Science*, 14, 441-443. 1983.

98. Habeebullah, A., Reghupathy, A., Rethinam, P. and Koshy, P.K., Residues of aldicarb in coconut. *Seminar on Pesticides and Environment*, Abstr. 37, TNAU, Coimbatore, 1983.

99. Sundararaju, P. and Koshy, P.K., Control of *Radopholus similis* on arecanut seedlings with aldicarb, aldicarb sulfone, carbofuran and fensulfothion. *Indian J. Nematol.*, 16, 4-7, 1986.

100. Sundararaju, P. and Koshy, P,K. Effect of different nematicides and *neem* oil cake in the control of *Radopholus similis* in yellow leaf disease affected arecanut palms. *Indian J. Nematol.*, 16, 44-47, 1986.

101. Venkitesan, T.S. and Setty, K.G.H., Control of burrowing nematode, *Radopholus similis* (Cobb) on black pepper. *Pesticides*, 13, 40-43, 1979.

102. Venkitesan, T.S. and Charles, J.S., A note on the chemical control of nematodes infesting pepper vines in Kerala. *Proc. Second Ann. Symp. on Plantation crops* (Placrosym II) p. 27-30, 1979.

103. Ramana, K.V., Slow wilt disease of black pepper and the role of plant parasitic nematodes in its etiology. Procedings of the Second Group Discussion on the Nematological problems of Plantation Crops at Central Coffee Research Institute, Chickmagalur, Karnataka, India, pp. 17-21, 1986.

104. Heungens, A., Nematode control in *Calathea* culture. *Medy. Fac. Landbouww. Rijksuniv. Gent.*, 47, 715-719, 1971.

105. Loos, C.A., Eradication of the burrowing nematode, *Radopholus similis* from bananas. *Plant Dis. Reptr.*, 45, 457-461, 1961.

106. Colbran, R.C., Effects of treatments of banana 'bits' for nematode control on emergence and yield. *Qd. J. agric. Sci.*, 21, 237-238, 1964.

107. Taylor, A.L., Control of the banana root nematode in Fiji. *FAO Pl. Prot. Bull.*, 17, 97-103, 1969.

108. Colbran, R.C. and Saunders, G.W., Nematode root-rot of bananas. *Qd. agric. J.*, 87, 22-24, 1961.

109. Percira, H.F., de Figueiredo, Jr. E.R. and Hussni, J., Nematoide cavernicola nos bananais do litoral de Sao Paulo, *Biologico* (Sao Paulo), 26, 27-31, 1960.

110. Senanayake, A.W., Root and rhizome necrosis black head disease of banana. *Trop. Agric.*, (Ceylon) 125, 119-121, 1969.

111. Blake, C.D., Root and corn diseases of bananas. *Agric. Gaz. N.S W.*, 74, 526-531, 533, 1963.

112. Blake, C.D., *Radopholus* root rot of bananas. *Commonwealth Mycological Conference* (6th), Lond. 1960. Report, pp., 53-54, 1961.

113. Mallamaire, A., La pourriture vermiculaire de bananier de Chine causee par *Anguillulina similis* Goodey, en Afrique Occidentale Francaise. *Agron. colon.*, 28, 33-42, 1939.

114. Feder, W.A., Ford, H.W., Feldmesser, J., Gardner, F.E.. Suit, R.F., Pieringer, A. and Hutchins, P.C., Citrus varieties, species and relatives susceptible to attack and damage by the burrowing nematode, *Radopholus similis*. *Plant Dis. Reptr.*, 42, 934-937, 1958.

115. Ford, H.W. and Feder, W.A., Three citrus rootstocks recommended for trial in spreading decline areas. *Circ. Fla Univ. agric. expl. Stn* No. 5-151, 8 pp., 1964.

116. Hutchinson, D.J., O'Bannon, J.H., Grimm, G.R. and Bridges, G.D., Reaction of selected citrus rootstocks to foot rot, burrowing and citrus nematodes. *Proc. Fla St. Hort. Soc.*, 85, 39-43, 1972.

117. O'Bannon, J.H. and Ford, H.W., An evaluation of several *Radopholus similis* resistant or tolerant citrus rootstocks. *Plant Dis. Reptr.*, 60, 620-624, 1976.

118. Shepherd, K., Banana breeding in the West Indies. *Pest News Summ. Soc.*, B 14, 370-379, 1968.

16

Genus Tylenchorhynchus Cobb, 1913

The stunt nematodes, earlier characterised under one genus, *Tylenchorhynchus* are now split into eight more genera namely, *Bitylenchus*, *Dolichorhynchus*, *Quinisulcius*, *Merlinius*, *Amplimerlinius*, *Uliginotylenchus*, *Triversus* and *Trilineelus*. Most of the information, however, on their association with diseased conditions of plants is under the former combined genus, *Tylenchorhynchus*. Therefore, for the purposes of this chapter, all the informations are being presented under the old names, with indications of the present status of their nomenclature. These are most commonly encountered nematodes in field soils and some species have been demonstrated to be positively involved with disease problems of economically important crop plants.

Morphology and Biology

The nematode body (about 1 mm long) is vermiform with slight tapering towards the extremities. The lip region is variable. In some species, there may be a constriction in the body cuticle, setting the lip region off from the body contour. The oesophagus is typically tylenchoid with a well developed vulvated median bulb. The glandular basal bulb does not overlap the intestine. Vulva is near middle of body. Females have paired and outstretched ovaries. Males are generally present with tails enveloped with large bursa.

In the recent times, the taxonomical status of the erstwhile genus *Tylenchorhynchus* has changed considerably with the number of lateral incisures as one of main distinguishing features. In the present concept, the earlier recognised genus has been split up into 9 genera. Though, earlier literature pertaining to this group refers to genus *Tylenchorhynchus* only, the informations presented here also deal with the earlier concept of the group with the new taxonomic status indicated wherever necessary.

Members of this group are ectoparasites, rarely encountered in the cortical cells, one to two cell deep, and feed on roots and root hairs as well. As the common name indicates, the plants infected with these species exhibit stunting, chlorosis, sparseness of leaf growth, etc.—general symptoms which can be encountered with other groups too. Root stunting may be seen also in some cases as a result of inhibition of cell elongation at feeding sites.[1] The life cycle (egg to egg), depending upon the species, generally may take 25-38 days. In the absence of host plants, some species have the ability to survive even upto ten months.[2]

Tobacco stunt nematode (*Tylenchorhynchus claytoni*)

It is one of the earliest species of the genus to have been demonstrated as pathogenic and as the main cause of stunting in tobacco fields of U.S.A. Subsequently, it was also found associated with diseased conditions in wheat, oats, corn, azaleas, etc. On tobacco, the infested roots exhibit shrivelling with sparse growth.[3] It has been quite frequently found associated with decline of azaleas in U.S.A.[4] Infested *Azalea indica* plants exhibit chlorosis, unthrifty growth and poor root system[5]. More damage occurs on 6 months old cuttings than the older ones[6]. Sometimes severe stunting and necrosis of roots can be seen also[7]. The species is also important on long leaf pine seedlings where it causes reduction in root weight[8].

Tylenchorhynchus dubius (*Bitylencus dubius*)

It is probably the most abundant polyphagous species in the lighter soils in the tempera areas of Europe[9]. It has been recorded *palustris*[15] and many others. Growth of potato plants has been directly correlated with the numbers of *T. dubius* present in soil.

The species thrives ectoparasitically feeding on roots of higher plants. On *Agrostis palustris*, it feeds on root hairs and the epidermal cells behind the meristematic zone. The life cycle period of the nematode is 40 48 days at 20°C.

Tylenchorhynchus brassicae

The species is presently known to be mainly a parasite of crucferous plants with cauliflower and cabbage as important hosts. On these hosts, it causes severe damage to new seedlings and in the older plants there is growth reduction. The root growth is suppressed and as a result, water absorption is affected[16]. The most favourable temperature for reproduction is 30°C and the most favourable soil moisture for its development is 30 per cent[17].

Tylenchorhynchus martini (*Tylenchorhynchus annulatus*)

It was first recorded on sugarcane in 1953[18] and later it was recorded to be widely associated with rice[19]. It feeds on root surface, root hairs and epidermal cells of sugarcane[20]. Inspite of the fact that it occurs rather in large numbers around sugarcane roots, it can, at best, be considered as a mild parasite only. The infested roots exhibit sparseness and are stubby. Under high infestation levels, the green weight of tops and roots is reduced and measureable damage may occur. Bermuda grass is another important host for the species in U.S.A. where its infestation has been recorded to affect the root development[21].

A swarming phenomenon, in this species, has been recorded. The swarming individuals cause more damage to sugarcane growth than observed with non-swarming individuals. It has been suggested that because of the swarming habit, the individuals clumped together around the roots and caused necrosis, which ultimately inhibited nutrient uptake[22].

The species can survive for quite a long period of time, without apparent feeding, in fallow soil. Under greenhouse conditions, it takes 14-21 days (egg to egg) to complete its life cycle on sugarcane roots.

Although the species has been recorded associated with plants, quite frequently and over large areas, still there is not much experimental work to evaluate its role in disease syndrome, if any. In India and Bangladesh, significant yield losses, in deep water rice, wherever these nematodes were present in the rhizosphere, have been recorded[23,24].

Tylenchorhynchus vulgaris (*Bitylenchus vulgaris*)

The species was first reported from around roots of maize[25] and was later established as a pathogenic species at a level of 1 nematode per g of soil[26] (Fig. 1). Most of the cultivated varieties of maize exhibit poor growth when inoculated with the nematode. It is mainly associated with graminaceous plants including grasses[27].

Fig. 1. Poor growth of maize seedlings after infestation by *Tylenchorhynchus vulgaris*.

A—Healthy seedlings ; B—Infested seedlings.

It has been recorded associated also with silvi-pastoral system of forage production[28]. Gram has been recorded also as a host for the species[29].

The nematode takes 25-27 days, at 25-30°C, to complete its life cycle on maize. However, the life cycle is considerably shortened (15-18 days) in the presence of *Hoplolaimus indicus* and *Pratylenchus thornei* in soil[30]. It is reported to multiply best in sandy loam soils with a pH range of 5.5-6.5[29].

Members of the family Graminae (including grasses) are reported to be the most suitable hosts for the species[27,29]. The species is also associated with silvi-pastoral system of forage production in *Cenchrus ciliaris* sown pastures[28]. Reports on the pathogenicity of the nematode are mainly confined to *Pennisetum typhoides, Zea mays* and *Cenchrus ciliaris.* A population level of 2 larvae per g of soil has been shown to be pathogenic for *P typhoides* and *C. ciliaris*, with percentage reduction in plant growth in *C. ciliaris* ranging from 15-17[31]. In several cases, the species, in combinations with other nematode species like *Meloidogyne incognita, Heterodera zeae* or *Pratylenchus zeae*, causes greater reductions in plant growth than alone. Whenever it was associated with *M. incognita*, it caused a reduction in gall formation but enhanced root penetration by larvae[32,33]. These reports are based on laboratory evaluations only. Any economic damage to field crops, under field conditions, has not been recorded as yet.

Tylenchorhynchus agri.

The species has been reported mainly as a parasite of leguminous and graminaceous plants[34]. On *Trifolium pratense* and *Poa pratense*, it causes growth reductions when the atmospheric temperature ranges between 23-26°C[35].

Some other species of *Tylenchorhynchus* recorded associated with economically important crops are: *T. microdorus* (*Merlinius microdorus*) on wheat, barley and oats; *T. capitatus* (*Quinisulcius capitatus*) on wheat and potato; *T. brevidens* (*Merlinius brevidens*) on wheat and barley; *T. curvus* (*Quinisulcius curvus*) and *T. crassicaudatus* with sugarcane. In all these cases, stunting and yellowing of the plants are the general above ground symptoms associated with high populations of these species. *T.* (*M.*) *brevidens* reduces grain yields in wheat and barley[36]

Interactions with Other Pathogens

A number of reports have been made, mainly based on laboratory studies, on the association of the stunt nematodes, with other micro-organisms. In these cases, either the disease situation gets aggravated or presence of one organism predisposes plant susceptibility for the other organism. Activity of a stunt nematode population associated with gram was observed to increase in the presence of the wilt fungus[37]. On the contrary, presence of *T. dubius*, around the rhizosphere of *Poa pratensis*, predisposed the plants to fungal (*Fusarium roseum*) attack[38]. A similar relationship has been reported in cotton with increased susceptibility of cotton to wilt in the presence of *T. claytoni*[39]. However, in an another case, *T. claytoni* did not affect wilting of plants but there was a greater multiplication of the nematode in the presence of the fungus[40].

Another interesting relationship has been recorded with *T. brassicae* and mosaic virus on brinjal. The virus promoted population build up of the nematode and it was suggested that because of the viral infection, certain physico-chemical changes occurred in plants which favoured nematode multiplication[41]. A synergistic effect on damage has been recorded also on cauliflower when *T. brassicae* is present alongwith *Rhizoctonia solani*, a root rotting fungus[42].

Apart from the interactions involving fungi and viruses, there are reports also about different nematode communities interacting, if present in the same niche, in the host-parasite relationships. In combination with *Meloidogyne hapla* or *Pratylenchus penetrans*, *T. agri*. was highly pathogenic on red clover.[43] Similarly, *T. brassicae*, which is rated as weakly pathogenic on tomato, caused significant growth reduction in combination with *M. incognita*[44]. These examples do not, however, signify that in combinations, any changes may be occurring in the behaviour of any of the nematode species. Increased damage may be the result of combined infestations only. Contrary to the enhanced effect of parasitism in combination with other nematode species, situations have been recorded also where presence of species of *Tylenchorhynchus* in the rhizosphere of plants have inhibited activities of other pathogenic species. Exposure of tobacco plants to *T. claytoni* has been recorded to inhibit invasion of roots by *Pratylenchus penetrans*[45]. Such interactions are also possible within different species of same or allied genera. On wheat both *Tylenchorhynchus* (*Merlinius*) *brevidens* and *T. vulgaris* multiply well singly but when they are in combination, then multiplication of *T. vulgaris* (*Bitylenchus vulgaris*) is reduced in presence of the former species[46].

REFERENCES

1. Wen, G.Y., Biology of the stunt nematode, *Tylenchorhynchu* species on soybean plants. *J. Alabama Acad. Sci.*, 40, 148, 1969.

2. Krusberg, L.R., Investigations on the life cycle, reproduction, feeding habits and host range of *Tylenchorhynchus claytoni. Nematologica*, 4, 187-197, 1959.

3. Graham, T.W., The tobacco stunt nematode in South Carolina. *Phytopathology*, 44, 332, 1954.

4. Barker, K.R. and Worf, G., Parasitism of 'southern stock' azaleas in Wisconsin by *Tylenchorhynchus claytoni*, *Meloidogyne incognita* and *Trichodorous christei. Phytopathology*, 54, 887, 1964.

5. Heungns, A., Chemische bestrijding van bodemmoeheid in de azaleateelt. *Meded. LandbHoogisch. Opzoekstns, Gent.*, 30, 1444-1453, 1965.

6. Aycock, R. and Doughtery, B.Z., Major diseases. In Growing Azaleas Commercially. (Eds. Kofranek, A.M. and Larson, R.A.), Univ. Califl, Davis, 78-88, 1975.

7. Radewald, J.D., Maire, R.G., Shibuya, F. and Sher, S.A., Control of stunt nematode (*Tylenchorhynchus claytoni*) on azalea (*Rhododendron indicum*). *Plant Dis. Reptr.*, 57, 540-543, 1972.

8. Ruehle, J.L., Influence of plant parasitic nematodes on long leaf pine seedings. *J. Nematol.*, 5, 7-9, 1973.

9. Sharma, R.D., Studies on the plant parasitic nematode *Tylenchorhynchus dubius. Meded. LandbHoogesch., Wageningen*, 71-1, 1-154, 1971.

10. Reynolds, H.W. and Evans, M.M., The stylet nematode, *Tylenchorhynchus dubius*, a root parasite of economic importance in the South West. *Plant Dis. Reptr.*, 37, 540-544, 1953

11 Jimnez, T.A., *Tylenchorhynchus dubius* Y *Pratylenchus neglectus* causa de declinacion in les Claveles (*Dianthus caryophyllus*) granadinos. *Revta. iber. Parasit.*, 26, 369-375, 1966.

12. Anonymous, Planteaygdomme i Denmark, 1962. *Tidsskrift. for planteavl.*, 67, 553-619, 1963.

13. Whitehead, A.G. and Fraser, J.E., Injury to field beans (*Vicia faba*) by *Tylenchorhynchus dubius. Plant Pathology*, 21, 112-113, 1972.

14. Kyrou, N.C., *Tylenchorhynchus dubius*, an injurious ectoparasite of potatoes. *Eur. Potato J.*, 12, 215-218, 1969.

15. Laughlin, C.W. and Vargas, J.M. Jr., Pathogenic potential of *Tylenchorhynchus dubius* on selected turf grass. *J. Nematol.*, 4, 277-280, 1972.

16. Alam, M.M. and Saxena, S.K., Effect of *Tylenchorhynchus brassicae* on the water absorption capacity of roots. *Geobios*, 2, 161-162, 1975.

17. Khan, A.M., Studies on plant parasitic nematodes associated with vegetable crops in Uttar Pradesh. *Final Tech. Rept., Grant no. F6-In-225, Proj. No. A 7-CR-65, Aligarh Muslim Univ., Aligarh, India*, 1969.

18. Birchfield, W., Parasitic nematodes associated with diseased roots of sugarcane. *Plant Dis. Reptr.*, 37, 38, 1953.

19. Fielding, M.J., *Tylenchorhynchus martini*, a new nematode species found in the sugarcane and rice fields of Louisiana and Texas. *Proc. Helm. Soc. Wash.*, 23, 47-48, 1956.

20. Birchfield, W. and Martin, W.J., Pathogenicity on sugarcane and host plant studies of a species of *Tylenchorhynchus. Phytopathology*, 46, 285-289, 1956.

21. Johnson, A.W., Pathogenicity and interaction of three nematode species on six Bermuda grasses. *J. Nematol.*, 2, 36-41, 1970.

22. Joshi, M.M. and Hollis, J.P., Pathogenicity of *Tylenchorhynchus martini* swarmers to rice. *Nematologica*, 22, 123-124, 1976.

23. Israel, P., Rao, Y.S. and Rao, V N., Rice parasitic nematodes. *Proc. Int. Rice Com., FAO, Lake Charles, Louisiana*, 1965.

24. Page, S.L. and Bridge, J., Root and soil parasitic nematodes of deep water rice areas in Bangladesh. *Int. Rice Res. Newsletter*, 4, 10, 1979.

25. Upadhyay, K.D., Swarup, G. and Sethi, C.L., *Tylenchorhynchus vulgaris* sp. n., associated with maize roots in India, with notes on its embryology and life history. *Indian J. Nematol.*, 2, 129-138, 1972

26. Upadhyay, K.D. and Swarup, G., Growth of wheat in the presence of *Merlinius brevidens* singly and in combination with *Tylenchorhynchus vulgaris. Indian J. Nematol.*, 11, 42-46, 1981.

27. Vaishnav, M.U. and Sethi, C.L., Reaction of some graminaceous plants to *Meloidogyne incognita* and *Tylenchorhynchus vulgaris. Indian J. Nematol.*, 7, 176-177, 1977.

28. Azmi, M.I. and Ahmad, S.T., Nematodes of silvi-pastoral system. I. Plant parastic and soil inhabiting nematodes associated with range grasses, pasture legumes and fodder trees. *Forage Res.*, 6, 105, 1980.

29. Upadhyay, K.D. and Swarup, G., Culturing, host range and factors affecting multiplication of *Tylenchorhynchus vulgaris* on maize. *Indian J. Nematol.*, 2, 139-145, 1972.

30. Siyanand, Seshadri, A.R. and Dasgupta, D.R., Investigations on

the life cycles of *Tylenchorhynchus vulgaris, Pratylenchus thornei, Hoplolaimus indicus,* individually and in combined infestations in maize. *Indian J. Nematol.*, 12, 272-275, 1982.

31. Azmi, M.I. and Singh. A., Pathogenicity of *Tylenchorhynchus vulgaris* on *anjan* grass (*Cenchrus ciliaris*). *Indian J. Nematol.*, 14, 184-185, 1984.

32. Vaishnav, M.U. and Sethi, C.L., Pathogenicity of *Meloidogyne incognita* and *Tylenchorhynchus vulgaris* on *bajra* and their interrelationships. *Indian J. Nematol.*, 8, 1-8, 1978.

33. Kaul, R.K. and Sethi, C.L., Effect of simultaneous inoculations and prior establishment of *Heterodera zeae, Meloidogyne incognita* and *Tylenchorhynchus vulgaris* on penetration of cyst and root knot nematodes into maize roots. *Indian J. Nematol.*, 12, 79-85, 1982.

34. Coates-Beckford, P.L. and Malek, R B., Host preferences of the stunt nematode, *Tylenchorhynchus agri. Plant Dis. Reptr.*, 62, 793-796, 1978.

35. Coates-Beckford, P.L., Influence of temperature and initial population density on population development and pathogenicity of *Tylenchorhynchus agri.* on *Trifolium pratense* and *Poa pratensis. Nematologica*, 12, 15-20, 1982.

36. Langdon, K.R., Struble, F.B. and Young, H.C. Jr., Stunt of small grains, a new disease caused by the nematode, *Tylenchorhynchus brevidens. Plant Dis. Reptr.*, 45, 248-252, 1961.

37. Sobun, N., Nema, K.G. and Dave, G.S., The possible interrelationship between plant parasitic nematodes (*Tylenchorhynchus*) and a root rot fungus from gram (*Cicer arietinum*). In Physiology of Parasitism (Eds. Aggarwal, G.P. and Bilgrami, K.S.). *Current Trends in Life Sciences*, 7, 451-456, 1979.

38. Vargas, J.M. Jr. and Laughlin, C.W., The role of *Tylenchorhynchus dubius* in the development of *Fusarium blight* of merion Kentucky bluegrass. *Phytopathology*, 62, 1311-1314, 1972.

39. Holdeman, Q.L., The effect of the tobacco stunt nematode on the incidence of *Fusarium* wilt in blue cured tobacco. *Phytopathology*, 46, 129, 1956.

40. Davis, R.A. and Jenkins, W.R., Effects of *Meloidogyne* species and *Tylenchorhynchus claytoni* on pea wilt incited by *Fusarium oxysporum* f. *pisi* race 1. *Phytopathology*, 53, 745, 1963.

41. Naqvi, S.Q.A. and Alam, M.M., Influence of brinjal mosaic virus in the population of *Tylenchorhynchus brassicae* and *Rotylenchulus reniformis* around eggplant roots. *Geobios* 2, 120-121, 1975.

42. Khan, A.M., Saxena, S.K. and Khan, M.W., Interaction of *Rhizoctonia solani* and *Tylenchorhynchus brassicae* in pre-emergence damping off of cauliflower seedlings. *Indian J. Nematol.*, 1, 8)-86, 1971.

43. Abosu, J.O. and Taylor, D.P., Stimulation of growth of red clover by *Tylenchorhynchus agri*. *Indian J. Nematol.*, 4, 132-137, 1974.

44. Alam, M.M., Masood, A. and Hussain, S.I., Effect of margosa and marigold root exudates on mortality and larval hatch of certain nematodes. *Indian J. Exptl. Biol.*, 13, 412-414, 1975.

45. Miller, P.M. and McIntyre, J.L., *Tylenchorhynchus claytoni* feeding on tobacco roots inhabits entry of *Pratylenchus penetrans*. *J. Nematol.*, 7, 327, 1975.

46. Upadhyay, K.D. and Swarup, G., Growth of maize plants in the presence of *Tylenchorhynchus vulgaris*, singly and in combination with *Pratylenchus zeae* and *Fusarium moniliforme*. *Indian J. Nematol.*, 11, 29-34, 1981.

17

Genus Hoplolaimus Daday, 1905

The genus was first recorded in 1905 from an island in the Paraguay river[1]. Despite its common occurrence and association with many cultivated crops of economic importance, throughout the continents, the role of lance nematodes, as potential pathogens, has not been fully explored. Pathogenicity of only few species has been established so far.

Morphology

These are large sized animals (1-2 mm) with set off, sclerotized lip region and having prominent transverse and longitudinal striae. The spear is massive with characteristic tulip-shaped basal knobs. Oesophageal glands are large and overlap the intestine, dorsally and laterally, with either 3 or 6 nuclei. The phasmids are large, one anterior and the other posterior to vulva. The female tail is hemispherical and annulated. There are 29 species recognised under the genus with only three, *H. columbus* Sher, 1963 ; *H. galeatus* (Cobb, 1913) Thorne, 1935 and *H. indicus* Sher, 1963 recognised as of some economic importance.

Distribution and Host Range

H. galeatus is the most common species in U.S.A., distributed in the Atlantic Coast states, Mississipi river basin, mid-west to Wisconsin and several region of south-western and western states.[2] *H. columbus* is known to occur in upper and middle coastal plains of south-

Carolina, Georgia, south western Georgia and North Carolina[2]. Both these species are known to occur in oriental regions and appear to be serious pest to cotton in Egypt. *H. indicus* was first described from soil around the roots of sugarcane in Punjab, India[3]. Subsequent studies indicate its widespread occurrence throughout India[4]. This species is common and second only to *Meloidogyne* in West Bengal (India), where it was associated with the failure of rice crop in two successive years[5]. *H. aegypti* and *H. columbus* are known to cause damage in cotton crop throughout Egypt[6,7]. *H. seinhorsti* and several other species are present in Africa[8–10]. Generally, *Hoplolaimus* species are cosmopolitan in distribution.

There is hardly any information regarding host range. Generally, woody and graminaceous plants support large populations of lance nematodes. Most of the reports indicate mere association of the nematodes with plant species rather than their preference for host. *H. galeatus* is associated with grasses, lespedeza, alfalfa, sugarcane, corn, hairy vetch, clover and cotton[2]. Many forest trees including species of *Pinus* are reported also to be of hosts of *H. galeatus*. It has been suggested that it may be responsible for serious losses to many forest trees[11,12]. It feeds and multiplies on rye, barley, wheat, oats, soybean, corn, cabbage and bean[13]. Pepper, eggplant, lettuce, tobacco and pea support limited population of the nematode. Host plants for *H. columbus* include Bermuda grass, wheat, field corn, millet, soybean, lima bean, cotton and sweet corn[14,15]. *H. columbus* is also known to parasitize sugarcane in Louisiana[16]. *H. indicus* has host preference for *Sorghum vulgare*, corn, *Citrus* spp., sugarcane, rice, jute and many grasses[4,5]. *H. aegypti*, originally reported from Egypt, is considered to be a serious pest on cotton. *H. stephanus* and *H. tylenchiformis* are found to be associated with ornamentals, grasses and shade trees[2].

Biology

Despite widespread occurrence of lance nematodes throughout the world, life cyccle studies have been conducted for only two species, *H. indicus*[4] and *H. columbus*[17].

In *H. indicus*, prior feeding is necessary for oviposition and development of successive post-embryonic stages. Within 24 hours of feeding, the female starts ovipositing. The expulsion of eggs between vaginal walls is brought about by intermittent and simultaneous contraction of ovary and uterus. The time interval between laying of first and second eggs varies from 15 minutes to 3 hours. A female oviposits on an average 14 eggs during a period of 4 days. Cleavage is completed by 48 hours and the first stage larva is observed within this period. The first moult occurs within the egg and development outside the egg consists of three larval stages and adult stage with three usual moults outside the egg. The life cycle, from egg to egg stage, is completed within 27-36 days at a temperature range of 28-32°C. Reproduction is amphimictic.

The life cycle of *H. columbus* is almost similar to *H. indicus* except for reproduction which is through mitotic parthenogenesis. The females lay 14-25 eggs lasting about 36 days and the interval between egg deposit and adult stage is about 45 days. Under field conditions, hosts, moisture and initial level of nematode population determine the population dynamics.[2] Highest population of *H. columbus* and *H. galeatus* is usually recorded at the harvest time and is correlated with the growing and active roots but is not influenced by rainfall. Population of *H. galeatus* declines after a dry period. Sandy loam soil is better than loamy sand and silty loam for reproduction[18].

Host-Parasite Relationships

The species have a wide range of feeding habit from migratory to endo-. semi-endo or even as ectoparasites. The feeding sequences, as observed in *H. indicus* and *H. columbus* essentially involve steps like exploration, stylet penetration, ingestion and stylet retraction[4,17,18]. The nematode movement towards root region appears to be direction oriented responding to stimuli emanating from host. All kinds of root may be parasitized. However, some species may attack only primary roots[2,17]. Usually *H. columbus* and *H. galeatus* continue to feed on the same root till they are completely embedded inside the host. Feeding may continue at one site for one to 9 days.[19] *H. galeatus* is considered of economic importance to cotton where it may cause damage rather severely. A population of 263 per 100 cm^3 in soil results in stunting, chlorosis and near defoliation of cotton plants[19]. The older roots are usually damaged more severely than the younger ones. However, it does not affect germination of cotton seed. Cotton plants may suffer considerable injury if the nematodes parasitise the roots endoparasitically. However, there are reports also that even an initial population of 4000 *H. galeatus* per 500 g soil does not do any harm to the growth of upland cotton. A population of 17000 to 25000 per 500 g soil may cause considerable injury to cotton[20].

The species is also responsible for economic losses in forest nurseries which may often exceed 50 per cent in terms of seedling mortality[11,12,21]. The nematode-infected seedlings show stunted and poor growth. Among the nursery plants, *Pinus* species often suffer serious losses[21]. Indeed this species is considered more destructive to loblolly pine (*P. taeda*) than any other nematodes. The infected roots exhibit dark colour with extensive damage to cortical tissues. The development of mycorrhiza, essentially beneficial factor in forest nurseries, is severely affected.

The species also infects many grass species and is considered to be pathogenic to turf grasses[22]. Growth of Bermuda grass var. Tifton 328 is singificantly reduced and roots of Kentucky blue grass (*Poa pratensis*) exhibit discolouration and complete destruction of roots[23]. In other grasses, there may be poor growth of top part as well as root resulting in chlorotic and retarded appearance of lawns. Magnesium and water uptake and translocation are significantly reduced in infected plants.

Besides cotton, forest trees and grasses, *Phalaris* sp. and *Festuca aurundinacea* are also parasitised by the species. Experimentally, it has been demonstrated that soil fumigation causes a drastic reduction in the nematode population and results in an yield increase of over 103 times more than non-fumigated plots[24,25].

H. columbus also causes significant losses in cotton. An initial population of 300-600 nematodes per 100 cm^3 soil suppress the root weight, yield and vigour of cotton cultivar Delta pine 16 and Coker 310[15]. In Egypt, where it is quite commonly encountered in cotton, peanuts, soybean and sesamum fields, it is reported to cause significant yield reductions[7,26]. An initial population of 700-800 females per pot results in 50 per cent loss in peanut yield[26]. Reduced yield of seed cotton, poor vegetative growth of the infected crop and lower number of bolls are often encountered in crops raised in heavily infested fields[7]. The critical damage level for peanut var. Giza 25, is estimated to be 700 females ; for cotton 400 females ; and only 300 females for *Sesamum indicum*.

H. aegypti is another species often recorded in cotton fields of Egypt. It has been estimated that the nematode causes a yield loss ranging between 15.6-53.3 per cent. *H. indicus* is suspected to inflict economic injury to cotton, rice and sorghum[4].

Formation of feeding cavities, cells without cytoplasm, thickening of cell wall and formation of membrane across the penetration point are some of the characteristic histological changes observed in infected cortical tissues of roots. In endodermis, the injury is often confined to 2-4 cells in all directions. In cotton roots, it often penetrates stele and feeds on phloem parenchyma and phloem elements. Consequently, such activities lead to abnormal division of phloem parenchyma, disorganization of cellular elements and eventual death of cells. Although the nematode is not reported to feed on xylem but its activities result in extensive damage to xylem. Tyloses are known to occur in damaged xylem cells.

In comparison to *H. galeatus*, cellular alterations by *H. columbus* in cotton are of lesser magnitude. Unlike *H. galeatus*, *H columbus* has no special preference for phloem tissues. Xylem vessels are also not injured. In soybean, it penetrates deep inside the root endodermis with the affected cortical tissues being ruptured and characterized in having vacuolated cells. Further, it prefers emerging rootlets for infestation. Extensive damage to cortex leads to sloughing off of cortical tissues which eventually results in limited number of secondary roots.

As a result of their migratory habit as well as ability to cause extensive tunnels in stele, lance nematodes provide avenues for the entry of other organisms in infected roots. *H. galeatus* predisposes cotton seedlings to the infection by *Fusarium* sp. *H. indicus* and *F. moniliforme* are associated with poor and patchy growth of sugarcane[27]. Similarly, *H. seinhorsti* and *F. oxysporum* are considered to be involved in a disease complex[28]. In most of such cases, the combined damage is much more than the individual effects. For instance, it has been demonstrated that with a population level of 17,000 nematodes/500 g soil, the damping off disease of cotton, attributed to *Rhizoctonia solani*, is significantly increased over the fungus alone effect[20]. Similarly, *H. columbus* in combination with *Pythium irregulare* causes much more damage to cotton than the individual effects of the organisms[29].

Control

Chemical application of 1,2-dibromo-3 chloropropane, ethylene dibromide, phenamiphos and carbofuran provide effective control against lance nematodes. Fumigants appear to be more effective than systemic and contact chemicals. A combination of early plantings and varietal selection in cotton are considered useful for reducing nematode populations[30].

REFERENCES

1. Thorne, G., Principles of Nematology. McGraw Hill Publications, 553 pp., 1961.

2. Lewis, S.A. and Fassuliotis, G., Lance nematode, *Hoplolaimus* spp. in the Southern United States. In Nematology in the Southern Region of the United States. So. Coopt. Ser. Bull. (Ed. Riggs, R.D.), 276 pp., 1982.

3. Siddiqi, M.R., Tylenchida: Parasites of Plant and Insects. CAB Publication, U.K., 645 pp, 1986.

4. Dasgupta, D.R., Siya Nand and Seshadri, A.R., Culturing, embryology and life history studies in the lance nematode, *Hoplolaimus indicus*. *Nematologica*, 16, 235-248, 1970.

5. Banerji, S.N. and Banerji, D.K., Occurrence of the nematod *Hoplolaimus indicus* in West Bengal. *Current Sci.*, 23, 597-598, 1966.

6. Shafiee, M.F. and Koura, F., *Hoplolaimus aegypti* n.sp. (Hoplolaimide : Tylenchida : Nematoda) from U.A.R. *Zool. Soc. Egypt*, 22, 117-120, 1974.

7. Koura, F. and Osmamtt, A., The relation between population density of *Hoplolaimus columbus* and damage to cotton and soybean. *Ann. Agri. Sci. Ann. Shaw Univ.*, 29 : 521-529, 1984.

8. Bridge, J., *Hoplolaimus seinhorsti*, an endoparasitic nematode of cowpea in Nigeria. *Plant Dis. Reptr.*, 57, 798-799, 1973.

9. Van den Berg, E., *Hoplolaimus seinhorsti.* C.I.H. Descriptions. Plant Parasitic Nematodes, Set 6, No. 6-7, 1976.

10. Van den Berg and Henys, J., South African Hoplolaiminae, the genus *Hoplolaimus* Daday, 1905. *Phytophylactica*, 2, 221-226, 1970.

11. Ruehle, J.L., Histopathological studies on pine roots infected with lance nematode and pine cyst nematodes. *Phytopathology*, 52, 68-71, 1962.

12. Ruehle, J.L., Plant parasitic nematodes associated with southern hardwood and coniferous forest trees. *Plant Dis. Reptr.*, 52, 837-839, 1964.

13. Ahmad, M. and Chen, T.A., Effect of certain environmental factors, and host plants on reproduction of *Hoplolaimus galeatus*. *Pl. Disease*, 94, 479-480, 1980.

14. Fassuliotis, G., Host range of Columbia lance nematode, *Hoplolaimus columbus*. *Plant Dis. Reptr.*, 58, 1000-1002, 1974.

15. Lewis, S.A. and Smith, F.H., Host plant distribution and ecological association of *Hoplolaimus columbus* on cotton and soybean. *J. Nematol.*, 8, 264-270, 1976.

16. Astudillo, G.E. and Birchfield, W., Pathology of *Hoplolaimus columbus* on sugarcane. *Phytopathology*, 70, 565 (Abstr.), 1980.

17. Fassuliotis, G., Feeding, egg laying and embryology of the

Columbia lance nematode, *Hoplolaimus columbus. J. Nematol.*, 7, 152-158, 1975.

18. Krusberg, L.R. and Sasser, J.N., Host-parasite relationships of the lance nematode in cotton roots. *Phytopathology*, 46, 505-510, 1956.

19. Lewis, S.A., Smith, F.H. and Powell, W.M., Host parasite relationships of *Hoplolaimus columbus* on cotton and soybean. *J, Nematol.*, 8, 141, 145, 1976.

20. Brodie, B.B. and Cooper, W.E., Pathogenecity of certain plant parasitic nematodes on upland cotton seedlings. *Phytopathology*, 54, 1019-1022, 1964.

21. Ruehle, J.L. and Sasser, J.N., The role of plant parasitic nematodes in stunting of pines in southern plantation. *Phytopathology*, 52, 56-58, 1962.

22. Perry, V.G., Smart, G.C. Jr. and Horn, G.C., Nematode problems on turf grasses in Florida and their control. *Proc. Fla. Stat. Hort. Soc.*, 83, 489-492, 1971.

23. Lukens, R.J. and Miller, P.M., Injury to turf grasses by *Tylenchorhynchus dubius* and *Hoplolaimus* spp. *Phytopathology*, 63, 240 (Abstr.), 1973.

24. Hoveland, C.S. and Haaland, R.L. and Rodriguez Kabana, R., Forage production of *Phalaris* species as affected by nematode population. *Nematropica*, 9, 22-27, 1979.

25. Rodriguez-Kabana, R., Haaland, R.L., Elkins, C.B. and Hoveland, C.S., Relationship of root-diameter and number of roots to damage of tall fescue by lance nematodes. *J. Nematol.*, 10, 297 (Abstr.), 1978.

26. Koura, F.H. and Osman, H.A., Growth response of peanut and sesamum as influenced by different inoculum levels, with lance nematodes, *Hoplolaimus columbus. Ann. Agri. Sci. Ann. Sham Univ.*, 29, 503-511, 1984.

27. Nath, R.P., Singh, R.K. and Haider, N.G., Combined effect of *Fusarium moniliforme* and *Hoplolaimus indicus* on sugarcane plants. *Sugarcane Pathol. Newslet.*, 17, 24-25, 1976.

28. Shammugam, N.A., Vinayagomurthi, A., Rajendran, G., Muthukrishnan, T.S. and Kandaswamy, T.K., Occurrence of nematode-fungus interaction on cotton around Coimbatore. *Sci. Cult.*, 43, 346-348, 1977.

29. Roncadori, R.W., Lehman, P.S. and McCarter, S.M., Effect of *Phythium irregulare* on cotton growth and yield and joint action with other soil borne pathogens. *Phytapathology*, 64, 1303-1306, 1974.

30. Nyczerpir, A.P. and Lewis, S.A., Relative tolerance of selected soybean cultivars to *Hoplolaimus* and possible effects of soil temperature. *J. Nematol.*, 11, 27-31, 1979.

18

Genus Paratylenchus Micoletzky, 1922

These are one of the smallest group of plant parasitic nematodes ranging between 0.25 to 0.5 mm. Because of the small size, the common name "Pin nematodes" has been ascribed to these animals.

Females are veriform but some mature females may be swollen in the region anterior to vulva. In the males, stylet is either degenerate or absent.

These nematodes are primarily ectoparasites, feeding for the most parts on epidermal cells and root hairs. They have been reported also to be present within roots, one to two cell deep in cortex. There are about 60 recorded species in the genus with about four or five species only about which limited informations on biology and pathogenicity are available. Under certain conditions, some of the species have been considered of some economic significance.

Hosts

These nematodes have been recorded from a number of plants and trees including mint, carnation, parsley, celery, lucerne, plum, fig, rose, etc.

Under field and natural conditions, large number of these nematodes have been encountered frequently around herbaceous and perennial crops. Limited informations on the life cycle indicate that eggs may be laid in clusters, with 7 days intervening between egg laying to hatching and 16 days from hatching to final moult[1]. In the case of *P. projectus*, the complete life cycle, from egg to egg, on *Festuca* sp. takes 36-38 days at 18-20°C[2]. In Italy, *P. dianthus* is reported to remain active throughout the year and reproduction was stimulated by low temperatures. Highest percentages of adults as well as the larval stages were recorded in winter. Males and the fourth stage larvae are reported as the most resistant stages and these stages do not feed on host tissues[3].

Symptoms

The symptoms of damage, as recorded, are mainly based on pathogenicity experiments. Some field observations have been made in cases of mint and carnations. Stunting, browning and rotting of subterranean stems of mint are recorded due to *P curvitatus*[4]. *P. hamatus* is known to cause rust brown colouration of celery roots[5]. On *Prunus* species, *P. neoamblycephalus* infestation results in dark as well as small roots having only few feeder roots[6]. In carnations, *P. dianthus* causes yellowing of leaves followed by reduced shoot and root growth[7].

Under experimental conditions, it has been established that some species can be quite damaging to the crops. *P. projectus*, at 2-7 nematodes/kg soil, can cause considerable yield reductions in red clover[8]. Similarly, *P. hamatus* seriously affects flower production in *Rosas hybrida*[9].

Apart from halogenated fumigants and granular chemicals which have been evaluated and found effective, some grass species like *Festuca arundinacea* and *F. rubra* have been found to reduce populations of *P. projectus*[10].

REFERENCES

1. Brezeski, M.W., Zepp, A.L. and D'Errico, F.P., Developmental stages of *Paratylenchus bukowinensis*. *Roczniki Nauk Rolniczych*, Seria E, 5, 145-151, 1976.

2. Wood, F. H., Biology and host range of *Paratylenchus projectus* from a sub-alpine tussock grassland. *New Zealand J. Agril Res.*, 16, 381-384, 1973.

3. D'Errico, F.P. and Di Maio, F., A three year study of seasonal dynamics and population density of *Paratylenchus dianthus* on carnation. In 'Nematodi delle Colture Ortofloricole e Industriat, Atti Giorante Nematologiche 1979, Firenze, Italy, 28-29, 1979.

4. Inagake, H., Furuyama, S., Kasano, H. and Sapurai, K., Studies on mint nematodes and their control. Res. Bull. Hokkaido Nat. Agr. Exp. Sta., No. 100, 48-57, 1972.

5. Brinkman, H., Nematological observations in 1975 and 1976. *Gewasbischeraning*, 8, 131-136, 1977.

6. Braun, A.L. and Lownsbery, B.F., The pin nematodes, *Paratylenchus neoamblycephalus* in myrobalan plum and other hosts. *J. Nematol.*, 7, 336-343, 1975.

7. D'Errico, F.P. and Di Maio, F., *Paratylenchus dianthus*, a parasite of carnation in Italy. *Informatone Fitopathologico*, 26, 5-6, 1976.

8. Townshend, J.L. and Potter, J.W., Forage legume yields and nematode population behaviour in microplots infested with *Paratylenchus projectus*. *Can J. Pl. Sci.*, 62, 95-100, 1982.

9. McDonald, D.H., Effects of *Paratylenchus hamatus* on productivity of greenhouse roses. *J. Nematol.*. 8, 294, 1976.

10. Townshend, J.L., Cline, R.A., Dirks, V.A. and Marks, C.F., Assessment of turf grass for the management of *Paratylenchus penetrans* and *Paratylenchus projectus* in orchards. *Can. J. Pl. Sci.* 64, 355-360, 1984.

19

Genera Hemicycliophora De Man, 1921 and Hemicriconemoides Chitwood and Birchfield, 1957

Members of these group of nematodes are ectoparasites. However, there is not much information about their economic importance. The common name "Sheath nematodes" for the members of the genus *Hemicycliophora* is because of an additional cuticle covering on the body of nematodes. The body cuticle as well as the sheath cuticle are produced at the same time. In the genus *Hemicriconemoides*, the outer cuticle is attached to the body at head, vulva and sometimes at tail tip. As per the latest taxonomical position of these two genera, although both the genera are placed in the sub-order Criconematinae, they have been put in separate families[1].

Genus *Hemicycliophora*

Members of this genus are quite common in several areas of the world but are most frequently encountered in temperate and sub-tropical regions. They have been recorded mostly around grasses and woody plants and less frequently around vegetable crops. Under experimental conditions, some species have been observed to parasitise large number of herbaceous annuals.

Generally, these are 0.4 to 2.0 mm long with rounded or truncate head. The stylet in females is well developed, 60 to 150 μm in length, with posteriorly sloped knobs. The males show degenerate stylet and have a characteristics semi-circular shape (sickle shaped) of spicules. Males also lack the sheath. There are about 100 species recognised under the genus.

As ectoparasites, they feed at or near the root tip. The stylet may penetrate several layers deep in the cortex while feeding. At feeding, some species, like *H. similis* and *H. arenaria*, have been recorded to produce adhesive plug circling the stylet[2]. The feeding period may be a prolonged one, as much as 15-30 minutes at a single feeding point or even several days. Life cycle studies with *H. arenaria* show that it takes 15-18 days at 28-30°C and some feeding was necessary between each moult[3].

Symptoms

Only few species like *H. typica*, *H. arenaria*, *H. similis*, etc. have received some attention with regard to their effects of parasitism on host plants. *H. typica* has been recorded as an important problem in carrots in The Netherlands where it causes production of short, stubby and malformed roots. *H. arenaria* has been considered of some importance to citrus trees. It causes cessation of root growth and initiates cell proliferation near the feeding site, resulting in formation of small galls. A number of females and larvae can be observed attached on a single gall. Tomato, pepper and celery are also parasitised by the species.

Unlike the root-knot nematodes, there is no giant cell formation in the galled roots. The cells which are penetrated by the stylet exhibit hypertrophied nucleus.

Genus *Hemicriconemoides*

Members of the genus *Hemicriconemoides* are world wide in distribution and include about 32 species. This genus was proposed for individuals which were intermediate to genera *Criconemoides* and *Hemicycliophora.*

Mature females have a sheath with plain annulation, 50-128 in number. The female stylet is quite robust with anteriorly cupped knobs. Males lack sheath and have a degenerate stylet.

The nematodes are commonly associated with turf and woody plants in tropical and sub-tropical regions. However, some species are known to parasitize sugarcane, sorghum as well as maize. High numbers of *H. chitwoodi* and *H. gaddi* have been recorded associated with stunted growth, tops as well as roots, of camellias. Under laboratory conditions, a species of *Hemicriconemoides* has been recorded to reduce significantly sugarcane growth at a level of 20 individuals/g soil[4]. *H. mangiferae* has been reported to be associated with severe decline of *Litchi chinesis*. The disease symptoms include presence of many bare twigs and branches, leaf chlorosis, leaf tip burn, poor flowering, excessive fruit drop and stubby root formation[5]. The nematode penetrates deep in the roots, often becoming completely embedded in the cortex. Under severe conditions of infestation, it also causes collapse of cortical tissues[6].

REFERENCES

1. Siddiqi, M.R., Tylenchida : Parasites of Plants and Insects. Commonwealth Agricl. Bur., Farnham Royal, U.K., 1986.
2. Kiesel, M., Castillo, J. and Zuckerman, B.M., An adhesive plug associated with the feeding of *Hemicycliophora similis* on cranberry. *J. Nematol.*, 3, 296-298, 1971.
3. Van Gundy, S.D., The life history of *Hemicycliophora arenaria* Raski (Nematoda : Criconematidae). *Proc. Helminth. Soc. Wash.*, 26, 67-72, 1959.
4. Sood, U., Studies on the genus *Hemicriconemoides* Chitwood and Birchfield, 1957. Ph.D. Thesis, Agra Univ., Agra, India, 1967.
5. Milne, D.L., Villiers, E.A. and Holtzhausen, L.C., Litchi tree decline caused by nematodes. *Phytophylactica*, 3, 37-44, 1971.
6. Milne, D.L., Appleton, M.R. and Holtzhausen, L.C., The anatomical reaction of *Litchi chinensis* roots attacked by three nematode species and a soil mite. *Phytophylactica*, 7, 15-20, 1975.

20

Genera Macroposthonia De Man, 1880, *Criconemoides* Taylor, 1936 and *Criconema* Hoffmanner and Menzel, 1914

The taxonomy of these nematodes has undergone drastic changes in the past, resulting in nomenclatorial revisions. Till these taxonomical changes, the important plant pathogenic genera recognised were species of *Criconemoides* and *Criconema*. Subsequently, two other important genera *Hemicriconemoides* and *Macroposthonia* were added to the group. Some of the species in the latter two genera are also considered important plant parasites. The common name 'ring nematodes' is mainly applied to members of genera *Criconemoides*, *Macroposthonia* and *Criconema*. Probably the common name 'ring' has been given to them because of coarsely striated cuticle which makes the nematodes appear nearly segmented giving an appearance of series of rings about the body or because of ring shape position, when they are relaxed by heat.

Genarally, the nematodes are short with 'sausage' shape body. The stylet is long with characteristic anchor shaped basal knobs. Another characteristic feature is the possession of thick retorse or rough or smooth prominent body annules. In the genus *Hemicriconemoides*, the body is enclosed in a cuticular sheath, attached to the body at head, vulva and sometimes tail tip.

The genus Criconema is recognised by its females lacking scales, spines and other cuticular structures, otherwise present in the juveniles. There is still some confusion relating to identity of the genera *Criconemoides* and *Macroposthonia* with the genus *Criconemella* but for the purposes of present presentation, the two genera are being treated separately since most of the literature deals under the previous nomenclature.

All these genera taken together are important ectoparasites which can do damage to plants when their soil populations are high. They are encountered frequently in sandy soils and soils with high pH.

Genus *Macroposthonia*

Amongst the over 80 species recorded under the genus, the best known species is *Macroposthonia xenoplax*. It is widely distributed in North America, Europe, South America, Africa, Australia, India and Japan. Host plants are generally woody but can attack also some herbaceous plants. The nematode was first reported in association with roots of Thompson seedless grape, walunt, plum, *Umbellularia californica* and pine[1]. Later, it was recorded around roots of *Pinus edulis*, *P. ponderosa*, *Juniperus deppeana*, *J. monosperma* and *J. scopularum*[2].

Maximum numbers of *M. xenoplax* are reported from California peach orchard during winter and early spring and lowest in late summer[3]. The reduction in numbers during summer was attributed to low moisture contents[4,5]. The nematode prefers sandy soils.

The life cycle is completed in 25-34 days[5]. Feeding is necessary by all juvenile stages for continuance of development. Feeding may take place at root tips or along roots, with no preference to any region. The life span of females is over two months.

Infested roots often do not show any obvious signs of damage but peach roots grown in soil, containing 50 nematode per g soil exhibit a shelf-like form[8]. Shoot and root growth and flower production have been reported to be reduced significantly in carnation by nematode infestation. This disease has been given the name 'ring nematode decline of carnation'[6]. Parasitism on Thompson seedless grape[7] and peach[8] has been demonstrated as well. The species is also known to be pathogenic to Japanese holly (*Ilex crenata*) where it produces interveinal chlorosis, leaf drop and stunted plants[9]. Another species, *M. onoensis* has been reported to be a problem in rice fields of Louisiana and Taxas (USA)[10]. The nematodes may be also partners in complex diseases. A root and stem rot disease of soybeans has been demonstrated to be aggravated when the nematode, *M. ornata* is present alongwith the fungus, *Phytophthora megasperma* var. *sojae*[11].

Preplant fumigation with DD at 40 gal/acre has been demonstrated to improve growth of peach replants[3].

The genera *Criconemoides* and *Criconema*, as now dilineated in the new proposals[12], have about a dozen species each which have been seldom found in large numbers in soils. Both are considered as ectoparasites but informations on pathogenicity or even parasitism are fragmentary.

REFERENCES

1. Raski, D.J., On the morphology of *Criconemoides* Taylor, 1936 with descriptions of six new species (Nematoda : Criconematidae). *Proc. Helminth. Soc. Wash.*, 19, 85-99, 1952.

2. Raski, D.J. and Riffle, J.W., Two species and further notes on *Criconemoides* Taylor, 1936 (Criconematidae : Nematoda). *Proc. Helminth. Soc. Wash.*, 34, 212-219, 1967.

3. Lownsbery, B.F., Studies of the nematode, *Criconemoides xenoplax*, on peach. *Plant Dis. Reptr.*, 43, 913-917, 1959.

4. Lownsbery, B.F., Factors affecting population levels of *Criconemoides xenoplax*. *Phytopathology*, 51, 101-103, 1961.

5. Seshadri, A.R., Investigations on the biology and life cycle of *Criconemoides xenoplax* Raski, 1952 (Nematoda : Criconematidae). *Nematologica*, 10, 540-562, 1965.

6. Sher, S.A., A disease of carnations caused by the nematode *Criconemoides xenoplax*. *Phytopathology*, 49, 761-763, 1959.

7. Raski, D.J. and Radewald, J.D., Reproduction and symptomatology of certain ectoparasitic nematodes on roots of Thompson seedless grapes. *Plant Dis. Reptr.*, 42, 941-943, 1958.

8. Thomas, H.A., Some initial findings on nematodes on peaches in New Jeresy. *Plant Dis. Reptr.*, 41, 526, 1957.

9. Aycock, R., Barker, K.R. and Bensen, D.M., Susceptibility of Japanese holly to *Criconeoides xemnoplax*, *Tylenchorhynchus claytoni* and certain other plant parasitic nematodes. *J. Nematol.*, 8, 26-31, 1976.

10. Hollis, J.P. and Keoboonrueng, S., Nematode parasites of rice. In Plant ant Insect Nematodes (Ed. Nickle, W.R.), New York and Basel : Marcel Dekker Inc., 95-146, 1984.

11. Campos, V.P. and Schmidt, D.P., The association of *Macroposthonia ornata* and *Phytophthora megasperma* var. *sojae* in the etiology of root and stem rot disease of soybean. *Summa Phytopathologica*, 7, 63-72, 1981.

12. Siddiqi, M.R. Tylenchida : Parasites of Plants and Insects. *Commonwealth Agril. Bur.*, Farnham Royal, U.K., 1986.

21

Genera Dolichodorus Cobb, 1914 and *Belonolaimus* Steiner, 1949

The genera *Dolichodorus* and *Belonolaimus* are long stylet bearing ectoparasites, mainly of grasses. The two well-known species of these genera, *D. heterocephalus* and *B. longicaudatus* have been recorded pathogenic on several field crops and therefore, the descriptions are mainly confined to these two species only. Their distribution is also mainly limited to the North and South America.

Genus *Dolichodorus*, The Awl Nematode

These are also commonly known as the 'awl' nematodes. The nematodes are characterised by having large offset labial disc, massive sclerotisation in the basal plate of cephalic framework, large stylet (54-170 μm) and well developed median and basal bulb. The female tail is conoid—spicate to filiform. The male tail, tapering to a fine acuminate point, possesses a large trilobed bursa.

Besides, *D. heterocephalus*, twelve other species are recognised but their potential as pathogens is not known.

Dolichodorus heterocephalus Cobb, 1914

It has been recorded principally in the USA, where it is mainly confined to the eastern states, particularly Florida. All records of severe damage to crops have so far come from this state. Although *D. heterocephalus* can be devastating where it occurs, outbreaks are localised[1].

D. heterocephalus is an ectoparasite, normally inhabiting wet locations such as swamps, marshes and the edges of lakes and streams. Under cultivated conditions, it occurs where soil moisture is high. Field infections in Florida may have originated through growers spreading infected soil from banks of the St. Jones river on their fields, and from use of irrigation water from lakes and ponds[2].

Males, females and all juvenile stages feed. Sometimes, individuals may remain at one location on the root upto 7 days. The nematode does not survive in fallowed pots after 3 months. There are indications of physiologic population also. For example, it has been seen that populations from Massachusetts decline on celery whereas the Florida populations multiply well on celery[3].

Severe stunting, consequent to depleted root systems of celery and corn has been observed in field infections in Florida[4]. The nematode seriously injures roots, reduces growth of plants and severely stunts water chestnut[5], celery, corn, tomato, bean and pepper[4]. The nematode feeds mostly at root tips but may feed also the sides of succulent roots, devitalising the tips, causing discolouration and necrosis. Large numbers of nematodes are seen in pit like lesions on hypocotyl of bean seedlings. Nematodes also prevent bean seeds to germinate by entering and feeding on the embryo[4]. Devitalisation of root tips has been observed quite often. Secondary roots, if produced, are also attacked. resulting in a much reduced stubby root type of symptoms. Lesions and necrosis are produced. In tomato, prolonged feeding results in the formation of brownish lesions around the feeding site and it results in cessation of root elongation. In parasitised roots, cortical nuclei enlarge and differentiated vascular tissue occurs nearer the tip than normal. Occasionally, well developed terminal galls are formed which show hypertrophy and hyperplasia of the cortical and vascular tissues.

Soil fumigants like DD or EDB have been reported to give excellent control of the nematode, under field conditions[2].

Genus *Belonolaimus*, the Sting Nematode

The best known species is *Belonolaimus longicaudatus*, which is considered a parasite of major importance in the south-eastern U.S.A. It is quite widespread throughout the Atlantic coastal plains, from Virginia to Florida.

It is a migratory ectoparasite, mainly restricted to light sandy soils. It is confined mainly to upper 30 cm soil but within this zone there is a seasonal as well as vertical fluctuation depending upon temperature. During high temperature conditions, prevailing in the month of July in Florida, greater population has been recorded below 15 cm soil layer than the top 15 cm. The reverse is true when the temperatures start declining from the month of September and onwards[6]. The nematode is favoured by high temperature and is most active between 20 and 34°C. Though the life cycle has not been investigated, it has been observed that all larval stages as well as adults have the capacity to feed on plant roots. In spite of its limited known distribution, three physiological races, with different host ranges, have been identified. Considerable differences in the host range between North Carolina and Georgia populations have been reported[7,8].

Plants attacked by *B. longicaudatus* exhibit severe stunting, leaf chlorosis, with a tendency to wilt in dry conditions. Crop yield is markedly reduced in quality as well as quantity. The nematode feeds ectoparasitically at the root tips and along the sides of the roots, resulting in a much reduced root system, often with sho stubby branches. Lesions and necrosis occur. In maize, gall like enlargements, at the end of the main roots, are produced by repeate killing of newly formed branches. On cotton, minute, dark, shrunken lesions are produced along the root axis or at the root tip in the early stages of attack. Lesions advance laterally to girdle off roots or advance length wise and persist without severing the roots. Histopathologically, the infested roots exhibit cavities that do not extend beyond the inner cortex. Cavities are large within, than their external opening, suggesting that the nematode, after locating a suitable feeding site, remains for some time merely moving its head. Feeding at the apical end suspends cell division at the tip, stimulating in other regions to produce the typical much branched root system[9].

The nematode has a wide variety of plant hosts such as cereals, grasses, forage crops, fruits, trees, ornamentals, vegetables and weeds. In Florida, the nematode is considered to be among the most important pests of turf and pasture grasses[10]. In some south-eastern cotton growing areas of U.S.A., *B. longicaudatus* has proved to be a devastating parasite, often killing all or most of the cotton plants in large areas of infested fields[11]. When it is associated with the *Fusarium* wilt fungus, almost total crop failure can result[12].

The sting nematode, in association with wilt fungi, causes severe losses in cotton. Resistant cotton variety, Cocker 100, becomes susceptible to wilt in the presence of the nematode[12]. The nematode has been recorded to facilitate the development of wilt in both susceptible and resistant cotton plants. Inoculations with *Fusarium* alone, does not cause any wilt in the resistant variety Cocker 100 and only 5 per cent wilting in a susceptible variety, but in association with *Fusarium* it produces 60 per cent wilt in Cocker 100 and 78 per cent in susceptible variety, after 70 days of seeding[4]. *B. longicaudatus* also interacts on Chrysanthemum with the fungus *Pythium aphanidermatum* where the symptoms appear sooner and in more severe form when both pathogens are together[13].

The sting nematode is quite susceptible to usual fumigants and organophosphates. Single applications are generally effective for only a single crop but in some instances protection of a second crop also has been realised. EDB (40%) at 10 gal/a has given a good control on groundnut[14] and at 20 gal/a on peppers and beans[15]. EDB, DBCP and other nematicides have been recorded also to suppress *B. longicaudatus* on strawberries throughout the season[14]. DBCP treatment has been recorded to increase yields of maize, groundnuts and soybeans by 100, 400 and 500 per cent respectively. Granular nematicides such as aldicarb, fensulfothion, carbofuran, cynem, lannate, phenamiphos and phorate have been found successful in the control of *B. longicaudatus* on maize, onion, cabbage, groundnut, potato, sorghum, soybean and cotton[16–18]. Greatly increased cotton yields have been recorded with mixtures of aldicar[illegible] ralin and pentachloronitro-benzene, consequent to reduction in population of *B. lonsicaudatus* and *Fusarium* wilt[19].

Cultural practices, like using *Crotolaria spectabilis* and *Tagetes minuta*, as cover crops, have given excellent control of sting nematodes[20]. Applications of organic nitrogen (activated sewage sludge) have reduced populations of *B. longicaudatus* on Bermuda grass and soils amended with alfalfa meal, cotton seed meal and rice straw have proved beneficial in reducing populations[21].

REFERENCES

1. Orton Williams, K.J., *Dolichdorous heterocephalus* Cobb, 1914 C.I.H. Descriptions of Plant Parasitic Nematodes, Set 4, No. 56, 4 pp., 1974.
2. Christie, J.R., Plant Nematodes : their Bionomics and Control. Gainesville, Fla. *University of Florida*, 1x+256 pp., 1959.
3. Paracer, S.M., Waseem, M. and Zuckerman, B.M., The biology and pathogenicity of the awl nematode, *Dolichodorus heterocephalus. Nematologica*, 13, 517-524, 1968.
4. Perry, V.G., The awl nematode, *Dolichodorus heterocephalus*, a devastating plant parasite. *Proc. Helminth. Soc. Wash.*, 20, 21-27, 1953.
5. Tarjan, A.C., Pathogenicity of some plant parasitic nematodes from Florida soils. III. Growth of Chinese water-chestnut, *Eleocharis dulcis* (Burm. f.) Henschel inoculated with *Dolichodorus heterocephalus* Cobb (Tylenchinae). *Proc. Helminth. Soc. Wash.*, 20, 94-96, 1953.
6. Boyd, F.T. and Perry, V.G., Effects of seasonal temperatures and certain cultural treatments on sting nematodes in forage grass. *Proc. Soil Crop Sci. Fla.*, 30, 360-365, 1971.
7. Abu-Gharbeih, W.I. and Perry, V.G., Host differences among Florida populations of *Belonolaimus longicaudatus* Rau. *J. Nematol.*, 2, 209-216, 1970.

8. Robbins, R.T. and Barker, K.R., Comparisons of host range and reproduction among populations of *Belonolaimus longicaudatus* from North Carolina and Georgia. *Plant Dis. Reptr.* 57, 750-754, 1973.

9. Standifer, M.S. and Perry, V.G., Some affects of sting and stubby root nematodes on graperuit roots. *Phytopathology*, 50, 152-156, 1960.

10. Orton Williams, K.J., *Belonolaimus longicaudatus*. C.I.H. Descriptions of plant parasitic nematodes, Set 3, (No. 40) 4 pp., 1974.

11. Sasser, J.N., Nematode diseases of cotton. In *Economic Nematology* (Ed. Webster, J.M.). Academic Press, London, 187-214, 1972.

12. Holdeman, Q.L. and Graham, T.W., The sting nematode breaks resistance to cotton wilt, *Phytopathology*, 43, 475, 1953.

13. Johnson, A.W. and Littrell, R.H., Pathogenicity of *Pythium aphanidermatum* to chrysanthemum in combined inoculations with *Belonolaimus longicaudatus* or *Meloidogyne incognita. J. Nematol.*, 2, 255-259, 1970.

14. Overman, A.J., Efficacy of certain nematicides in strawberry production. *Proc. Fla. St. Hort. Soc.*, 76, 114-119, 1964.

15. Christie, J.R., The sting nematode can be controlled by soil fumigation. *Down to Earth*, 9, 8-9, 1953.

16. Rhoades, H.L., Effect of nematicides on yield of field corn in central Florida. *Proc. Soil Crop Sci. Soc. Fla.*, 28, 262-265, 1969.

17. Rhoades, H.L., Effect of preplant and post-plant nematicides on yield of field corn attacked by sting nematodes. *Nematropica*, 1, 5, 1971.

18. Sasser, J.N. Nelson, L.A. and Barker, K.R., Effects of soybean following chemical soil treatment. *J. Nematol.*, 4, 233-234, 1972.

19. Brodie, B.B. and Hauser, E.W., Multiple pest control in cotton with mixtures of selective pesticides. *Phytopathology*, 60, 1609-1612, 1970.

20. Good, J.M., Minton, N.A. and Jawonski, C.A., Relative susceptibility of selected cover crops and coastal Bermuda grass to plant nematodes. *Phytopathology*, 55, 1026-1030, 1965.

21. Tomerlin, A.H. Jr., Ph.D. Dissertation. *Univ. Florida, Gainsville*, 49 pp., 1969.

22

Genus Aphelenchoides Fischer, 1894

The genus *Aphelenchoides* comprises large number of species with diverse habitats, wide host range and varied mode of parasitism. They are usually small slender nematodes, morphologically distinct from tylenchids in having dorsal gland orifice opening in the lumen of median bulb anterior to the valve, weak stylet, usually with inconspicuous knobs, monodelphic gonad ; males without gubernaculum and "rose thorn" shaped spicules. There are more than 137 nominal species that have been described and recorded, which include plant and insect parasites, fungivorous and predators. Amongst these, the most important group is bud and leaf nematodes which are considered as potential pathogens of higher plants. *A. besseyi*, causing white tip disease of rice and summer dwarf or crimp of strawberry ; *A. ritzemabosi* causing severe damage to chrysanthemum ; *A. fragariae* causing spring crimp of strawberry and *A. composticola*, a limiting factor in mushroom cultivation, are the most important species in the group. Two other species, *A. blastophthorus* and *A. bicaudatus* are considered also of economic importance but their role in crop damage has not been fully assessed. *A. bicaudatus* is soil-inhabiting mycophagous nematode, encountered in soil around roots of bamboo, corn, cotton, garlic, grapefruit, onion, potato, rice, sorghum, sugarcane, tomato, turf grass wild cherry and egg plant. It is suspected that feeding by this nematode on endomycorrhizal fungi restricts development of the beneficial fungi and interferes with the uptake of nutrients[1].

Both ecto- as well as endoparasitism are observed in the group. Ectoparasitism is frequent in plant crowns, leaf axils, or inflorescence where the nematode is well protected within folded plant parts. Where the host plants have succulent leaves like *Begonia*, endoparasitism is known to occur[2]. Although these species are parasites of higher plants, they can be also successfully cultured on fungi.

Biology and Symptoms

Amongst the plant parasitic nematodes, *Aphelenchoides* species are known to have the shortest duration of life cycle. Under optimum condition, these nematodes complete their life-cycle in one and a half weeks.

The pre-adult fourth-stage larvae is the infective stage in the life history of the nematodes[3]. *A. besseyi* survives as pre-adult beneath the hull of rice kernel. One infested rice seed may contain 5 to 14 pre-adult nematodes. When nematode-infested seeds are planted in the moist soil, the desiccated and quiescent fourth-stage larvae become active and move with growing rice plants, where they feed ectoparasitically. Under tropical conditions, two to three days after feeding ectoparasitically, the pre-adults moult to become males and females. The females lay eggs in the leaf axils and flower panicles. The first moult apparently occurs within egg. The second- and third moults take place in 3 and 1-2 days respectively. It takes about 6-8 days for the second-stage larvae to become females and males. At 21-23°C, the life cycle is completed in 8-10 days, and several generations may occur in a crop season[2,3]. The nematode fails to develop at a temperature below 13°C. Parthenogenetic as well as amphimictic types of reproduction have been reported for this nematode[4].

The temperature and humidity are two important factors that appear to influence the activities of *A. besseyi*. The quiescent pre-adults in rice grain can survive from 23 months[5,6] to 8 years[7]. Within a temperature range of 17.9-25.1°C, these nematodes can thrive well. Under experimental conditions, the pre-adults become quiescent in 4, 6 and 8 days at relative humidities of 30, 50-70 and 90-100 per cent, respectively[8].

A. ritzema-bosi lays 25-30 eggs in a compact group. The eggs hatch in 3-4 days and the second-stage larvae mature to adult in 9 to 10 days. In crysanthemum, this species completes one generation in 12-14 days[9,10]. The cryptobiotic nematodes survive 16 to 36 months in dried leaves. The species can withstand low temperatures between 4 to −2°C and can regain their activities[10]. Another species, *A. fragarlae* has almost identical life cycle. At 18°C, the eggs hatch in four days and in 6 to 7 days the second-stage larvae become adults.

Aphelenchoides besseyi Christie, 1942, The White Tip Nematode of Rice

It is prevalent throughout rice growing areas of the world. It also causes summer dwarf or crimp of strawberry in the United States of America and Australia. The disease has been also referred to as French bud, white bud, briar bud, possum ears, frenching, curly leaf, mole plant and wild plant[4].

It was first recorded in Japan where yield losses were estimated between 17.4 to 54.1 per cent. It is now reported from all parts of rice growing areas of the world[11]. In Thailand, yield losses have been recorded to vary between, 20 to 60 per cent[12] and about 40 per cent in India[13,14]. It is also suspected to cause economic losses in U.S.A. and Bangladesh though exact estimates are not available.

It is an ectoparasite, feeding on growing point of rice seedlings resulting in whitening of 3 to 5 cm of leaf tip which becomes necrotic. Other symptoms are crinkling and distortion of flag leaf (Fig. 1) enclosing the panicle, which may be obstructed from emerging, and reduction in the the size of panicle, number of grain and size of the grain. Under field conditions, the disease expression is not very prominent during the early stage of crop growth. In many instances, some plants exhibiting severe form of symptoms may recover subsequently and appear normal. Likewise, in boot leaf stage, the nematode-infected plants may not produce typical symptom of twisted flag leaf. The glumes of infected spikelets remain white and do not undergo any change, either in shape or size. Nematodes, in such spikelets, reside on the inner side of lemma and palea, at the base of floral parts. The nematodes feed

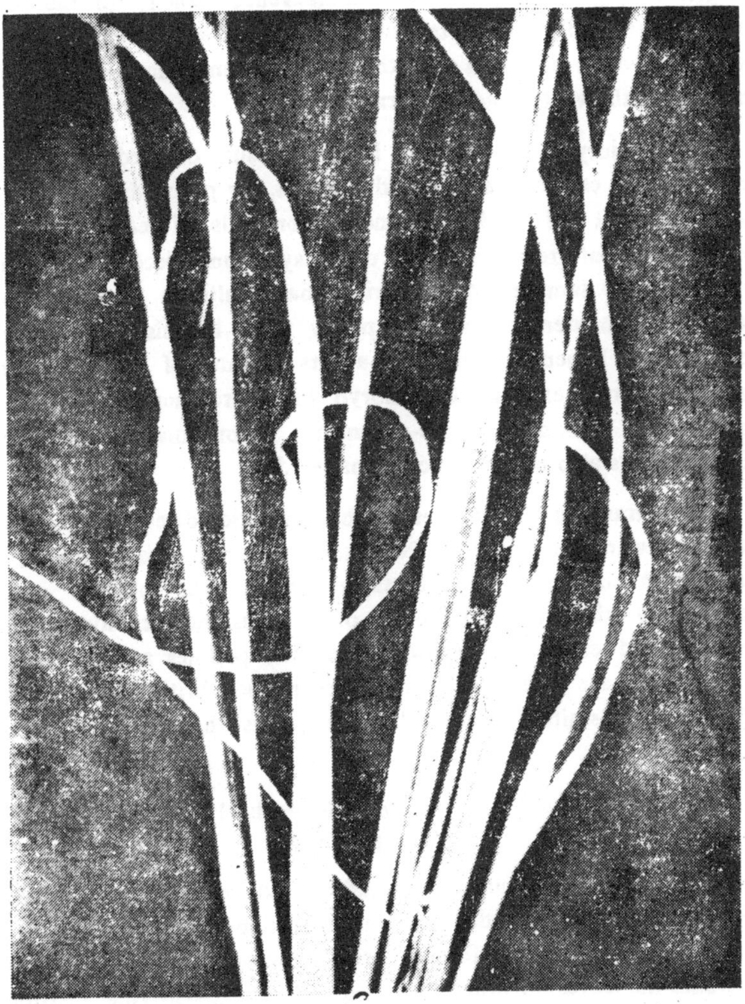

Fig. 1. Distortion of flag leaf of rice plants caused by *Aphelenchoides besseyi*.

externally on the lodicules and ovary, leading to shrivelled and empty grains giving an appearance of chaffiness[2,5,15,16].

Two types of symptoms are known to be associated with white tip disease of rice grains[17]. A damage during early milk stage results in longitudinally cracked and blackened kernels. In the late milk to early yellow stage, transversely cracked and wedge shaped kernels develop. In such a case, the symptoms on grains are observed mostly on the ventral side of kernels.

The injury to rice plants is as a result of feeding by the nematode on cells of leaf primordia, leaves, ear primordia and ears. Mis-alignment of epidermal and motor cells of leaf, chlorosis, bronzing of leaf are some of the other symptoms recorded in this disease. Cell damage to ear primordia results in emergence of earheads with sterile grains or partly filled kernels. The injury inflicted on lodicules, adversely affects closure of flowers after anthesis, thereby exposing the embryo to environmental stress and invasion to secondary micro-organisms. Abnormal elongation of floral parts is also observed occasionally[15,16].

In strawberry, it causes a characteristic disease known as summer dwarf or crimp[2,18,19]. Nematode infested leaves are deformed reduced in size, crinkled and asymmetrical. The infected young leaves may exhibit upward cupping of margins and the older leaves a downward tendency. In diseased plants, petioles are short and the leaflets are elongated with greener and britle leaves. The disease is transmitted to other daughter plants through the mother plant.

Aphelenchoides ritzema-bosi (Schwartz, 1911) Steiner and Buhrer, 1932

On chrysanthemum, the nematode feeds on buds and growing points resulting in retarded growth, distortion and deformation of leaves with roughened feeding areas[2]. Subsequently, the parasite invades the leaves on which they feed thereby destroying mesophyl cells of parenchymatous tissue, resulting in formation of leaf spots or blotches on the underside of leaves[20]. Gradually, these blotched areas turn brown and black, and as the infestation spreads, the diseased areas become limited by primary leaf veins giving characteristic angular appearance. With the death of the tissues, the nematodes migrate through the stomata, into terminal buds on which they continue to feed. This results into deformed and under-sized flowers. With the progression of the disease, it is possible to see varied symptom expressions in the same plant. Dead leaves, angular blotches, yellowing or bronzing between veins, blackened buds and outgrowth of side shoots with deformed flower buds are some of the common symptoms.

African violet and black current are some other plants which suffer most due to infection with *A. ritzema-bosi*. These plants show distorted and stunted growth, curly leaves, defoliation and in extreme case, death of infected hosts.

Aphelenchoides fragariae (Ritzema-Bos, 1891) Christie, 1932

It is a serious pest of strawberry. The nematode feeds as an ectoparasite within the folded crown, buds, leaflets and leaf axils. Occasionally, small, dry and brown feeding areas may be observed on expanded leaves, usually near the mid rib, indicating feeding activities of the nematode in the folded bud stage. The disease is generally characterized by reduced size of leaves that are crinkled, cupped, dark dreen, brittle, less pubescent, and reddish in colour[2,21].

In association with bacterium, *Corynebacterium fascians*, it causes a disease named as cauliflower disease of strawberry[2,22]. Initially, the diseased plants exhibit stunting with deformed stems, leaves and flowers. Enormous development of calyx, accompanied by enlarged sepals, small or lack of petals, greenish colour, are also some of the prominent symptoms expressed in this disease complex. With advancement of disease, the severely stunted plants resemble a small cauliflower.

The nematode also produces two other diseases by its ownself. The diseases are, red plant and spring dwarf or crimp of strawberry[2]. Deformed and reduced growth of buds, stems, leaves and flowers, crinkled leaves and reduced flowering are the common symptoms expressed in red plant disease. Few strawberry varieties like Royal Sovereign, Lexton's King George and Duke exhibit red symptoms of foliage.

The spring dwarf or crimp occurs during early spring in North Atlantic states. The symptoms are more or less identical to red plant disease.

A. fragariae may also act like endoparasite in the leaves of ferns, and begonia. The nematode infestation of these plants produces typical leaf blotch symptoms.

Aphelenchoides compositicola **Franklin, 1957**

This species is considered to be an important limiting factor in mushroom cultivation. The symptoms are rather atypical, non-specific and difficult to detect. Reduced mushroom yields give the first indication of the presence of the nematode (Fig. 2). Generally, a high first flush followed by poor second one, is an indication of the attack by the nematode. Sparse and patchy growth of mycelium, whiteness of spawn turning into brown, poor sporophore yield per flush and delayed flushes are some other recorded symptoms[23].

Aphelenchoides blastophthora **Franklin, 1952**

It causes destruction of inflorescence and distortion of leaves with reduction of laminae in *Scabious*[24]. In severe infestation, the normal flower is replaced by a small 'blind bud', otherwise the infested plants appear with deformed unevenly developed inflorescence.

Some species of the genus are involved also with fungi or bacteria where they may aggravate the symptom expression.

Acrocylindricum oryzae and *Corticium* species have been found to invade inter-veinal areas of *A. besseyi*-affected rice leaf to cause enhanced injury to crop[16]. Discolouration of rice kernels has been reported to be due to association of *A. besseyi* with *Enterobacter agglomerans*[17]. The classic example is that of cauliflower disease of strawberry which is expressed only when *A. frogariae* is associated with the bacterium, *Corynebacterium fascians*[22]. In this disease, the nematode probably functions as a vector by transporting the bacterium to the meristematic regions of the crown.

Distribution and Host Range

Amongst all the species, *A. besseyi* is the most widespread nematode throughout rice growing areas of the world. It has been recognised also as economically important, particularly to rice and strawberry. It has been recorded from Japan, India, Bangladesh,

Indonesia, Taiwan, USSR, Italy, Hungary, El Salvador, Central and West Africa, Cuba, Cameroon, U.S.A. *A. besseyi* has a narrow host range, mainly confined to rice and strawberry, *Polyanthes tuberosa Ficus elastica* and *Sporobolus poiretii*[25].

A. ritzema-bosi, commonly known as chrysanthemum foliar nematode, is also widespread in Europe and United States of America[2,21]. It parasitizes, ecto and endoparasitically, in the aerial parts of more than 247 plant species. Besides, chrysanthemum, its hosts include strawberry, black currant, dahlia, delphinium, african violet, gaillardia, phlox, primula, carnation, zinnia, annual aster and many weed plants[26,27]. It is an important parasite on fruits, like strawberry and black currant in Europe and U.S.A. The annual loss in chrysanthemum has been estimated to range between 45 to 92 per cent[28].

A fragariae is also a common and widespread species noted to occur in temperate and tropical zones of Europe and U.S.A. mainly under moist conditions. It is reported to parasitize more than 250 plant species belonging to 47 families. The strawberry decline in France is attributed to this nematode[29]. It is quite prevalent in strawberry growing area of Poland (more than 34 per cent of fields). The nematode accounts for 54 per cent loss in strawberry fruit yield[30]. Control of the nematode results in increase of crop yield by 63 to 82 per cent[31]. In addition to strawberry, other common hosts include fern, begonia, anemonae, coleus, primula, ranunculus, saintpaula, croft lily and violet.

This species can be cultured successfully on the fungus *Alternaria citri.*

A. compositicola is recorded from almost all mushroom growing areas of the world. It is a very serious pest of this crop in Europe and Asia. One nematode per 20 g of compost can completely destroy the crop[23].

Control

White Tip Disease

Hot Water Treatment

The disease is primarily dessiminated through infected grains, husks and plant remains. Soaking paddy seed in warm water at 55 to 61°C without pre-soaking or at 52 to 53°C with pre-soaking for 15 minutes will kill the nematodes, without adversely affecting seed germination[5,6]. Sun drying of seed for several days during summer also gives a satisfactory control of the nematode.

Resistance

Arkansas Fortuna, Nira 43, Bluebonnet, Improved bluebonnet, Centuary 53, Gurmatia, Chinoor, Co 13, Rohini IR. 5 and SR 26B are some of the promising rice varieties which exhibit resistance against the nematode[14,16,32].

Chemicals

Foliar spray of parathion and systox effectively controls the disease. A nematicide-fungicide compound, N-244, as a seed treatment, provides satifactory control of white tip when the chemical is applied at the rate of 4 oz per bushel, with a 10 per cent formulation[32].

Cultural

Seeding of rice directly in water reduces the disease incidence to 0.5 per cent as against to 60 per cent when the seeds are drilled or flooded. Burning of plant remains, after the harvest of crop, is desirable from the control point of view[33,34].

Summer Dwarf of Strawberry

Use of certified propogating plant material to provide disease free strawberry plants to growers is the best method to control the disease caused by *A. besseyi*[2].

Chrysanthemum Foliar Nematode

Hot Water Treatment

Immersion of chrysanthemum stools at 44·4°C for 30 minutes is advocated for the control of *A. ritzema-bosi*[35].

Chemicals

Foliar sprays of chrysanthemum plants with a weak solution (0.005 per cent) of parathion, applied twice, in monthly intervals, is effective for the control of the nematode[36]. A dip in 0.03 per cent parathion for 20 minutes of chrysanthemum stools also controls the nematode. In strawberries, a 5 minutes exposure to parathion, at 0.07 per cent, gives a good control.

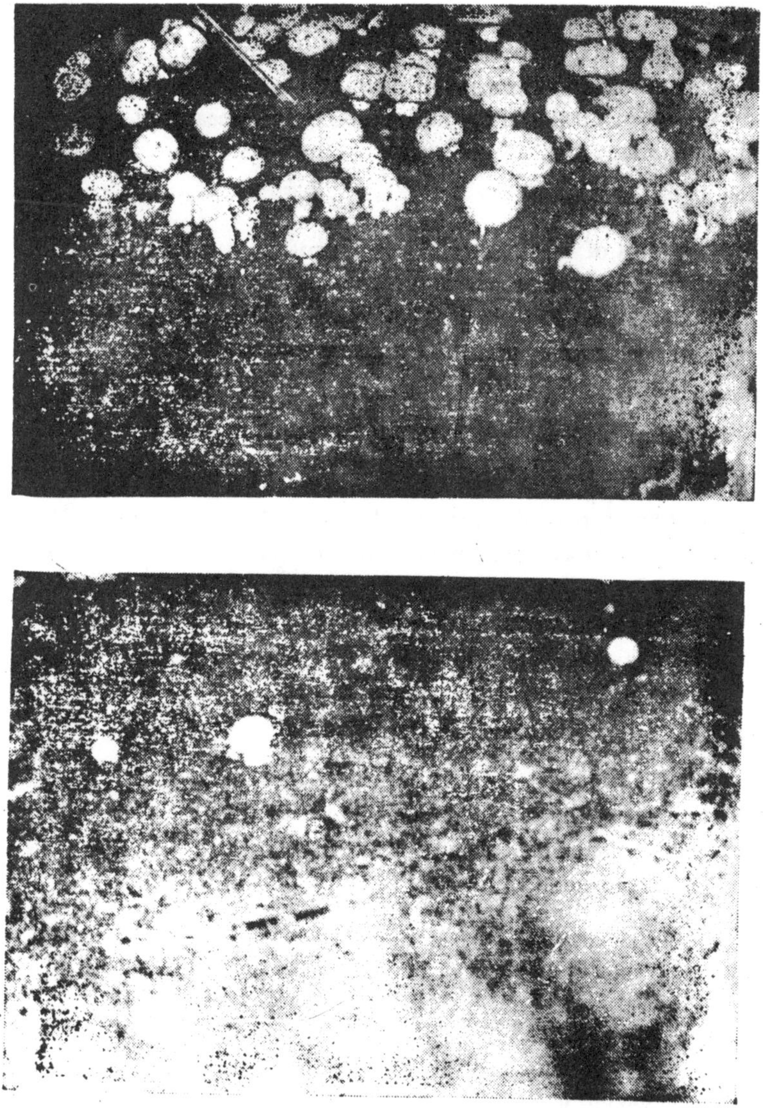

Fig. 2. Drastic reduction in mushroom growth by *Aphelenchoides compositicola*, in comparison to healthy (above).

Resistance

Chrysanthemum varieties, Any shoesmith, Orange Beauty and Orange Peach are reported to be resistant[37].

Cultural

Propogation of chrysanthemum through top cuttings reduces the degree of infestation considerably. Cleaning and burning of infested plant remains also provide a good control measure[38].

Spring Dwarf Disease

Hot Water Treatment

Treatment of strawberry runner plants at a temperature of 47°C for 15 minutes followed by a cold plunge in water controls *A. fragariae* causing spring dwarf disease of strawberry.

Resistance

Strawberry varieties, George Soltwedel, Regina and Talizman are reported to be resistant against *A. fragariae*.

Chemicals

Parathion sprays offer promise for control of the nematode. Immersion of strawberry plants in 0.2 per cent solution of thionazin 20 E or of isozophos 50 before planting in soil, infested with *A. fragariae*, keep the plants free of nematodes and increase fruit yield[39].

Control of Mushroom Nematode

Chemical

Chemical methods should be restricted to compost only. Thionazin is the only chemical recommended for the control of *A. compositicola* without any toxic residual effect.

Disinfection of trays and implements with formalin or with boiling water is an efficient method for the control of nematode.

REFERENCES

1. Shafer, S.R., Rhodes, L.H. and Riedel, R.M., *In vitro* parasitism of endomycorrhizal fungi of ericaceaus plants by the mycophagous nematode, *Aphelenchoides bicaudatus*. *Mycologia*, 73, 141-149, 1981.

2. Throne, G., Principles of Nematology. McGraw-Hill Book Company, Inc. N.Y., Toronto, Lond., 553 pp., 1961.

3. Sivakumar, C.V., Post-embryonic development of *Aphelenchoides besseyi in vitro* and its longevity in stored rice seeds. *Indian J. Nematol.*, 17, 147-148, 1987.

4. Huang, C.S. Huang, S.P. and Chiang, Y.C., Mode of reproduction and sex ratio of rice white tip nematode, *Aphelenchoides besseyi*. *Nematologica*, 25, 255-260, 1979.

5. Atkins, J.G. and Todd, E.H., White tip disease of rice. 1. Symptoms, laboratory culture of nematodes and pathogenicity tests. *Phytopathology*, 48, 632-637, 1958.

6. Todd, E.H. and Atkins, J.G., White tip disease of rice. II. Seed treatment studies. *Phytopathology*, 49, 184-188, 1959.

7. Zem, A.C., Momteira, A.R, and Bahia, J., The white tip nematode also occurs in rice seeds. *Revista de Agriculture*, Piracicaba, Brazil, 52, 81-82, 1977.

8. Rao, Y.S., Rao, J., Prasad, J.S. and Panwar, M.S.. Behaviour of the white tip nematode of rice in relation to temperature and humidity. *Oryza*, 20, 72-73, 1983.

9. Wallace, H.R., Observations on the behaviour of *Aphelenchoides ritzema-bosi* in chysathemum leaves. *Nematologica*, 5, 315-321, 1960.

10. French, N. and Barraclough, R.M., Survival of *Aphelenchoides ritzema-bosi* (Schwartz) in soil and dry leaves. *Nematologica*, 7, 309-316, 1982.

11. Atkins, J.G. and Todd, E.H., White tip disease of rice. III. Yield tests and varietal resistance. *Phytopathology*, 49, 189-191, 1959.

12. Taylor, A.L., Nematode parasites of rice. In Nematodes of Tropical Crops (Ed. Peachey, J.E.). Commonwealth Agriculture Bureau, St. Albans, U.K., 355 pp., 1969.

13. Muniappan, R. and Seshadri, A.R., On the occurence of white tip nematode of rice, *Aphelenchoides besseyi* Christie, 1942 in Madras State. *Madras Agri. J.*, 51, 510-511, 1964.

14. Dastur, J.F., A nematode disease of rice in central province. *Proc. Ind. Acad. Sci.*, 4,(2), 108-122, 1936.

15. Muthukrishnan, T.S., Rajendran, G. and Chandrasekharan, J., Studies on the white tip nematode of rice, *Aphelenchoides besseyi* in Tamil Nadu. *Indian J. Nematol.*,4 , 188-193, 1974.

16. Rao, Y.S., Prasad, J.S. and Panwar, M.S. Nematode problem in rice : Crop losses, symptomatology and management. In Plant Parasitic Nematodes of India : Problems and Progress, pp. 279-299, IARI Publication, 1986.

17. Nishizawa, T., Occurence of abnormal rice kernels associated with the infestation of white tip nematode, *Aphelenchoides besseyi. J. Nematol.*, 6, 73-79, 1976.

18. Christie, J.R., Spring dwarf and summer dwarf of strawberries. *U.S. Dept. Agri. Cir.*, 681, 1-10, 1943.

19. Christie, J.R., A description of *Aphelenchoides besseyi*, n.sp. the summer dwarf nematode of strawberries, with comments on *A. subtenuis* (Cobb) and *A. hodsoni* Goodey, 1935. *Proc Helminthol. Soc. Wash.*, 9. 82-84, 1942.

20. Stokes, D.E. and Alfieri Jr., S.A., A foliar nematode and *Phytophthora parasitica* on *Gloxinia*. *Fla. St. Hort. Soc.*, 81, 376-380, 1968.

21. Slack, D.A. and Riggs, R.D., The bud and leaf nematodes. In Nematology in Southern Region of the United States. *So. Coopt' Ser. Bull.*, 276, 179-182, 1982.

22. Crosse, J.E. and Pitcher, R.S., Studies on the relationship of eelworms and bacteria to certain plant diseases. I. The etiology of strawberry cauliflower disease. *Ann. appl. Biol.*, 39 475-484, 1952.

23. Sharma, N.K. and Seth, Anju, Nematode problems in Mushroom : Losses, symptomatology and management. In :

Plant Parasitic Nematodes of India : Problems and Progress (Eds. Swarup, G. and Dasgupta, D.R.), pp. 384-399, 1986.

24. Franklin, M.T., A disease of *Scabiosa caucasia* caused by the nematode, *Aphelenchoides blastophthorus* n.sp. *Ann. appl. Biol.*, 39, 54-66, 1952.

25. Marlatt, R.B., Relationship of disease methods of inoculation of *Ficus elastica* with *Aphelenchoides besssyi*. *Plant Dis. Reptr.*, 56, 471-472, 1972.

26. Brown, E.B., A seed borne attack of chrysanthemum eelworm (*Aphelenchoides ritzema-bosi*) on annual aster (*Callisptephus chinensis*). *J. Helminth.*, 30, 145-148, 1956.

27. Goodey, J.B., Franklin, M.T. and Hooper. D.J., The nematode parasites of plants catalogued under their host. (3rd Edn.). *Farnham Royal Comm. Agric. Bureau.*, 214 pp., 1965.

28. Baranowski, T., Investigations on the harmful influence on chrysanthemum grown in Pozanan and its neighbourhood. *Roezniki Nauk Rolniczych*, Seria E.6, 17-31, 1976.

29. Clerjeau. M. Rancillar, M. and Deschambre, D., The position regarding strawberry decline in France. Pe'pinierisles Horticulueurs Maraichers. *Rev. Horticol.* No. 237, 39-42, 1983.

30. Bohmer, B., The harmful effect of *Aphelenchoides fragariae* and *A. ritzema-bosi* on *Fragaria aranassa*. *Pflanzen*, 33, 113-117, 1981.

31. Tacconi, R., Trials on the control of epigeal nematodes of strawberry (*Aphelenchoides fragariae* and *Ditylenchus dipsaci*). *Fitopathologica*, 30, 3-8, 1980.

32. Cralley, E.M., A summary of recent investigations on the control of white tip of rice. *Rice Annual*, 19-20, 1953.

33. Cralley, E.M., A new control measure of white tip. *Ark. Farm. Res.*, 5 (4), 5, 1956.

34. Cralley, E.M., Effect of seeding methods on the severity of white tip of rice. *Phytopathology*, 47, 7 (Abstr.), 1957.

35. Christie, J.R., Plant nematodes, Their Bionomics and Control. *Fla. Agri. Expt., Stat.*, 256 pp., 1959.

36. Dimock, A.W. and Ford, C.H., Control of foliar nematode of chrysanthemum with parathion sprays. *Phytopathology*, 40, 7 (Abstr.), 1950.

37. Hesling, J.J. and Wallace, H.R., Observations on the biology of chrysanthemum eelworm, *Aphelenchoides ritzema-bosi* (Schwartz) Steiner in florist's chrysanthemum. *Annal. appl. Biol.*, 49, 195-203, 1961.

38. Christie, J.R., The influence of chrysanthemum propagating material on dissemination of the foliar nematode. *Proc. Helminth. Soc. Wash.*, 9, 1-4, 1942.

39. Tacconi, R., Control trial against foliar nematodes of strawberry (*Aphlenchoides fragariae* and *Ditylenchus dipsaci*). *Helminth. Abstr.*, 49 (4), 1979.

23

Rhadinaphelenchus cocophilus (Cobb, 1919) Goodey, 1960

The nematode is associated with the well-known and devastating disease, red ring of coconut. Though the disease was suspected to be present in coconut gardens of Trinidad Islands for quite some time, the first authentic technical reference appeared in 1905[1]. Later, the disease was recorded also in Grenada and it was confirmed that a nematode was the causal agent of the disease[2]. The nematode was later described under the name of *Aphelenchus cocophilus*[3] which was subsequently redesignated to a new genus *Rhadinaphelenchus cocophilus*[4]. It is a major problem facing the coconut industry in Carribean Islands and Central and South America. So far, it is known to be confined to these areas and has not been recorded from other coconut growing areas of Africa, South-East Asia, Pacific Islands, Florida or Cuba[5].

An insect, black palm weevil (*Rynchophorus palmarum*) has been found to be associated with the transmission of the disease to coconut palms[6]. While feeding on the palms, the insect picks up the nematode infested tissues and such insects when they feed on healthy palms, deposit these infested tissues in the bases of leaf axils. The nematode, thus deposited, can feed on apparently undamaged surface all types of coconut tissues. All stages of the insect are able to carry the nematode infestation, either internally or externally, for transmission. It has been demonstrated that 35-45 per cent of the weevils, collected from an infested area, carry living nematodes, with an average burden of 71 nematodes [7,8]. Live nematodes have been recovered from the digestive tract, body cavity as well as from the ovipositor region of the palm weevil[7]. Interestingly, live nematodes have been recovered also from the faeces of the weevil[9,10]. Large numbers of the nematode larvae have been recorded to enter the body cavity of the palm weevil larvae, either through the mouth or spiracles. The nematode persists, without change or multiplication, throughout the metamorphosis phases of the insect. On emergence of the adult stage of the insect, large number of nematodes, located in the region of ovipositor of the female, are injected into the soft tissue of the coconut palm, at the time of deposition of eggs by the insect. Such weevils, carrying the nematodes, are slightly smaller in size (30 mm long as against the population mean of 33.48 mm) and it has been suggested that this may be due to the effect of nematode on its development[11,12].

Although, palm weevil is an important factor in the spread of the disease, coconut trees can get infected through other mechanisms as well. It may be mentioned here that the nematodes can survive for quite a long time and can move through the soil for some distance[13]. The nematode can survive in fresh water films for 7-8 days and in sea water for 3 days[6], but in soil, it can survive upto 15 days[14]. In soil, the nematode movement is favoured by high moisture and infested coconut tissues lying in soil, near the trees, can initiate disease incidence. High degree of disease incidence has been recorded in swampy low lying areas[13,15]. Nematodes are capable of invading healthy or damaged or mature or young roots of palms. Following initial infection, the nematodes spread inwards along an individual root until the infection reaches the base of the palm. Within 10-30 weeks after infection, a generalised root infection apears outwards to a distance of several feet. The appearance of this infection is succeeded in 3-4 weeks by the onset of red ring symptoms and later by the death of the palm[15]. Such root infestations, however, do not always develop into a stem infection. The failure to establish a systemic infestation in a palm, following root infestation, indicates that root system of same palms possess a degree of resistance to the nematode. Observations on the symptomatology and severity of attacks suggest that Panama Tall is more resistant/tolerant to the nematode than Jamaica Tall[1].

Nematode infestation is known to kill palms. Palms of bearing age (5-7 years) are more susceptible to attack than the older or younger palms. The former group exhibits a high degree of resistance, although disease incidence has been recorded in palms, as old as 20 years.

The primary symptom of the disease is the appearance of yellowing of the lower leaves, starting at the tips of distal pinnae and extending progressively downwards to the base of the pinnae and leaves. The yellowing is succeeded by browning and eventual death of the leaf. The yellowing and browning subsequently spread to inner leaves and the trees die within 3-4 months after the appearance of the first symptom.

The internal condition of the stem of the palm suffering from red ring is very characteristic. In cross section, approximately 2.5 cm beneath the stem surface, an orange-red ring discolouration, about 3 cm in width, is seen. At first, scattered dots, about 1 mm in diameter, are seen which eventually coalesce and form the ring of reddened tissue. Hence the names red ring diseases and red ring nematode. The ring normally extends the whole length of the stem and roots and into the petioles where it takes on a crescent-like shape. Large numbers of juveniles are present in the discoloured tissue. Adult nematodes are usually located in the peripheries of the ring. As high as 50,000 nematodes in 10 g of infected coconut stem tissue has been reported[16]. Nematodes are also present in the root cortices which on infestation by nematodes, become spongy in texture and reddish brown in colour. Shedding of green nuts, of all sizes, takes place with or slightly in advance of the leaf discolouration and sometimes is seen as the first observable symptom. The advanced stages of infection may be associated with putrefaction of the soft tissues of the stem and base of central column of unexpanded leaves and sometimes the palm breaks off at the neck[5,13,17].

R. cocophilus also causes little leaf disease of coconut and oil palm in Surinam and Guyana[18]. The disease is characterised by leaves being reduced in size and standing stiffly upright; pinnae are short, wiry and necrotic terminally; yellowish suberised patches appear on leaf bases and petioles and older leaves turn yellow to grey. The disease has been reproduced experimentally in oil palm[19].

A minimum of 5,000 juvenile nematodes are required in the small area of leaf axil, to initiate disease[20]. If the contaminated weevil are caged on the inter nodal tissues, immediately above the leaf axil, then upto 72 per cent disease transmission can take place[21]. It has been speculated that some other insects of coconut palm may be involved also, in a minor role, in the transmission of the disease[22].

Normally, the nematodes do not invade the vascular tissues. However, in the upper stem portions, the nematodes invade the intercellular areas between the immature fibres forming the vascular sheath[14]. Water uptake of the infected coconut palms is markedly reduced as a result of occlusion of xylem vessels[14,23]. The nematode has one of the shortest life cycle, completing in 9 to 10 days[24,25].

The red ring disease is recognised as one of the most serious problems in coconut plantations. Thirty-five per cent mortality of young coconut palms has been recorded in Trinidad. In a single plantation in Tobago, as much as 80 per cent of the palm trees are reported to be destroyed by the disease[16]. In Venezuela, 35 per cent of oil palms have been destroyed by the disease within a period of 10 years[26].

Control

At present, no effective control measure is known. However, the disease incidence can be reduced with the application of chemicals, both against the weevils as well as the nematodes. Monthly application of 5 per cent granular nemaphos to leaf axils was found effective in checking the disease, but was not recommended as a practice because of high mamallian toxicity as well as treatment cost[14]. Apart from insecticides, which reduce the insect population, wire mesh baskets, filled with diseased coconut tissues, soaked in 0.1 per cent lannate solution, have been used for attracting the weevils in the basket. Apparently, the weevils are attracted towards diseased coconut tissues and get killed by the insecticide[26]. Destruction of infested palms has been suggested also as an integral part of plantation hygiene. The foul smell of the diseased palms attracts palm weevil which then get contaminated with nematodes or nematode infested debris may get stuck to their bodies. Therefore, wherever possible destruction of surviving diseased palms is a recommended practice[27,28].

Another method that has been tried to reduce nematode populations, is injection of nematicidal chemicals like phosdrin alongwith sodium arsenite, into stem of diseased trees. Injecting 30 ml sodium and potassium arsenite in downward sloping holes, near base of stem, has been found effective in killing 95 per cent of the nematode population in palms. It was also effective in killing palm weevil larvae. An additional advantage was that the treated trees dried very quickly so that ultimate burning and destruction of diseased trees became easy[29-31].

Most of the cultivated varieties are susceptible to the disease. However, some varieties like Carib in Venezuela ; Green Spanish in the Carribean ; Cockerite in Guyana ; Panama Tall in El Salvador ; and Manila cultivar in Ecuador, have been reported to be tolerant. Artificial inoculations by nematodes on young nuts results in development of disease symptoms. This method has been used for screening coconut germplasm and some resistance has been found in some nuts[32].

Although, a number of methods have been tried, singly or in combination and found effective in reducing the disease incidence, no single method is likely to limit its spread. Today, even after 70 years, our knowledge about the disease is meagre and we do not have anything better to offer to the coconut grower. The only effective and lasting solution to this notorious disease, occurring in several countries[33], is a resistant and high yielding variety which can only be achieved by a sustained breeding programme. Locating resistance in palms with good agronomic characters and multiplying them through tissue culture, though difficult to achieve, could be also an answer.

REFERENCES

1. Hart, J.H., Bud rot disease of coconuts, Gulf Coast. *Bull. Misc. Infn. Roy. Bot. Gardens* Trin., 6, 242-243, 1905.

2. Nowell, W., Root disease of the coconut palm in Grenada. *Agric. News, Bridgetown*, 17, 389-399, 1918.

3. Cobb, N.A., A newly discovered nematode, *Aphelenchus cocophilus* n.sp. connected with a serious disease of the coconut palm. *W. Indian Bull.*, 17, 203-210, 1919.

4. Goodey, J.B., *Rhadinaphelenchus cocophilus* (Cobb, 1919) n. comb., the nematode associated with 'red ring' disease of coconuts. *Nematologica*, 5, 98-102, 1960.

5. Dean, G.G., Red Ring Disease of *Cocoa nucifera* L. caused by *Rhadinaphelenchus cocophilus* (Cobb, 1919) Goodey, 1960. An annotated bibliography and review. *Tech : Commun. Commonw. Bur. Helminth.* No. 40, St. Albans, 1979.

6. Ashby, S.F., Some recent observations on red ring disease. *Agric. News, Bridgetown*, 20, 250-351, 1921.

7. Fenwick, D.W. and Mohammed, M., The presence of *Rhadinaphelenchus cocophilus* (Cobb, 1919) Goodey, 1960 on the body of the palm weevil, *Rhynchophorus palmarum* (L.). *A. Rep. Res. Stn. Trinidad & Tobago Coconut Research* Ltd., 1961, 1-15, 1962.

8. Hagley, E.A.C., The palm weevil *Rhynchophorus palmarum* L., a probable vector of red ring disease of coconuts. *Nature*, 193, 499, 1962.

9. Hagley, E.A.C., The role of the palm weevil, *Rhynchophorus palmarum* as vector of red ring disease of coconuts. 1. Results of preliminary investigations. *J. Econ. Ent.* 57, 375-380, 1963.

10. Tidman, D.A., Agricultural and horticultural problems in Brazil. *Wild Corps*, 3, 341-364, 1951.

11. Griffith, R., Progress on the entomological aspects of Red Ring Disease of coconuts. 1. Persisting *Rhadinaphelenchus*

cocophilus (Cobb, 1919) Goodey, 1960 during metamorphosis of the palm weevil *Rhynchophorus palmarum* related to their presence on the body surface of newly emerged insects. *J. agric. Soc. Trin.*, 67, 209-213, 1967.

12. Griffith, R., The relationship between the red ring nematode and the palm weevil. *J. agric, Soc. Trin.*, 68, 342-346, 1968.

13. Fenwick, D.W., Red ring disease of the coconut palm. In Nematodes of Tropical Crops (Ed. Peachey, J.E.) *Tech. Commun. Commonw. Bur. Helminth.* No. 40, 89-98, 1969.

14. Blair, G.P., Red ring disease of the coconut palm. *J. Agric. Soc. Trin.*, 64, 31-49, 1964.

15. Fenwick, D.W., Red ring disease of the coconut palm. In Tropical Nematology (Eds. Smart Jr., G.C. and Perry, V.G.). Gainesville, University of Florida Press, 38-48, 1968.

16. Esser, R.P., *Rhadinaphelenchus cocophilus*, a potential threat to Florida palms. *Nematology Circular, Division of Plant Industry, Fla. Dept. Agric.*, No. 9, 1969.

17. Brathwaite. C.W.D. and Siddiqi, M.R., *Rhadinaphelenchus cocophilus*. C.I.H. descriptions of plant parasitic nematodes, Set 5, No. 72, 1975.

18. Van Hoof, H.A. and Seinhorst, J.W., *Rhadinaphelenchus cocophilus* associated with little leaf of coconut and oil palm. *Tirdschr. PlZiekt.* 68, 251-256, 1962.

19. Maas, P.W.T., Contamination of the palm weevil (*Rhynchophorus palmarum*) with the red ring nematode (*Rhadinaphelenchus cocophilus*) in Surinam. *Nematologica*, 16, 429-433, 1970.

20. Griffith, R., The mechanism of transmission of the red ring nematode. *J. Agric. Soc. Trin.*, 68, 437-457, 1968.

21. Hagley, E.A.C., The mechanism of transmission of *Rhadinaphelenchus cocophilus* by the palm weevil, *Rhynchophorus palmarum*. *Phytopathology*, 55, 117-118, 1965.

22. Hagley, E.A.C., Role of insects as vectors of red ring disease. *Nature*, 204, 905-906, 1964.

23. Blair, G.P. and Darling, H.M., Red ring disease of the coconut palm, inoculation studies and histopathology. *Nematologica*, 14, 395-403, 1968.

24. Blair, G.P., The use of immature nuts of *Cocos nucifera* for studies on *Rhadinaphelenchus cocophilus*. *Nematologica*, 11, 590-592, 1966.

25. Blair, G.P., Studies on red ring disease of the coconut palm. *Oleagineux*, 25(1), 19-22, (2) 79-83, 1970.

26. Griffith, R., Red ring disease : The mechanism of spread and recommendations for control. *Nematropica*, 1, 2, 1971.

27. Nowell, W., Diseases of coconut. In : Diseases of Crop Plants in the Lesser Antilles. The West Indian Committee, London, 383 pp., 177-182, 1923.

28. Malaguti, G., 'Pudricion del cogollo' de la Palmera de Aceite Africana (*Elaeis guineensis* Jacq.) en Venezuela. *Agronomla trop.*, 3, 13-31, 1953.

29. Blair, G.P., The problem of control of Red Ring Disease. In Nematodes of Tropical Crops (Ed. Peachey, J.E.). *Tech. Commun. Commonw. Bur. Helminth.*, No. 40, 99-108, 1969.

30. Potter, T.E.K., Pamphlet on the control of the palm weevil *Rhynchophorus palmarum* L. with weedicide 100, Port of Spain, Trinidad : T. Geddes Grant, 1953.

31. Maharaj, S., The use of arsenic on coconut trees suffering from Red Ring Disease. *J. Agric. Sco. Trin.*, 64, 59-66 1964.

32. Blair, G.P., The use of immature nuts of *Cocos nucifera* for studies on *Rhadinaphelenchus cocophilus*. *Nematologica*, 11, 590-592, 1966.

33. Govindan Kutty, M.P. and Koshy, P.K., Nematodes associated with the coconut palm (*Cocos nucifera* L.). In Nematodes, Fungi, Insects and Mites Associated with the Coconut Palm. *Tech. Bull.* 2, 1-50, C.P.C.R.I., Kasaragod, Kerala, India, 1979.

24

Genera Xiphinema Cobb, 1913 ; Longidorus (Micoletzky, 1922) Filipjev, 1934 ; Paralongidorus Siddiqi, Hooper & Khan, 1963 ; Trichodorus Cobb, 1913 & Paratrichodorus Siddiqi, 1974

Nematode species, known to transmit plant viruses, belong to a single order, Dorylaimida. Even within this order, nematode vectors belong to a small group that are placed under five genera, *Xiphinema*, *Longidorus*, *Trichodorus*, *Paralongidorus* and *Paratrichodorus*, the latter two being relatively recent additions [1,2]. These genera received considerable attention after the discovery of pathogenic capability of the ectoparasitic nematodes[3], and virus vector capability of nematodes for the first time in 1958[4]. Besides transmitting virus diseases, these nematodes are parasites on several economically important agri-horticultural crops and forest plantations. Due to recent taxonomic revisions, these five genera have been split further into eighteen. However, this chapter is being confined mainly to these five genera only since there is practically no information on the diseases caused by them.

I. Genus *Xiphinema*, The Dagger Nematode

Xiphinema ensiculiferous was the first dagger nematode described as early as 1893 by Cobb[5], under the genus *Tylencholaimus*. However, *X. americanum* is the type species of the genus, as such the first species described under the generic name in 1913[6]. Since then, several other species have been described. Sixty eight valid species of the genus *Xiphinema*, known in 1975[7], rose to 112 by 1980[8], and atleast another 39 species have been added since then (Table 1), though some of them have been synonymised recently. Of the known species, *X. americanum*, *X. index* and *X. diversicaudatum* are of wide occurrence and are economically important. Several species have assumed economic significance only because of their ability to transmit virus diseases otherwise they are weak parasites. Majority of the species attack perennial fruit trees and forest plants, where pathogenicity proof and estimation of economic losses are difficult tasks.

TABLE 1

List of *Xiphinema* spp. described after 1980

Species	Reported country
(1)	(2)
algeriense	Algeria
bacaniboia	Fiji
barbercheckae	South Africa
barense	Italy
bolandium	South Africa
capense	South Africa
clavicaudatum	Brazil
*cobbi**	India
coomansi	South Africa
colombiense	U.K.
dolosum	Nigeria
fatikae	Nigeria
fluminense	Brazil

(Contd.)

(*Contd.*)

(1)	(2)
*hayati**	Algeria
hydrabadensis	India
incertum	Bulgaria
indica	India
kosaigudensis	India
limbeense	Malawi
loteni	South Africa
lusitanicum	Portugal
magaliesmontanum	South Africa
majus	Nigeria
malawiense	Malawi
*manmillocaudatus**	India
nagarjunensis	India
occiduum	Canada
oryzae	Nigeria
pachydermum	Portugal
papuanum	Papua New Guinea
pararadicicola	India
pseudocoxi	Germany
riocaguetae	U.K.
simile	Bulgaria
stenocephalum	Ivory Coast
stockeri	Botswana
thorneanum	U.S.A.
xenovariable	South Africa

*Now synonymised with known species.

These are long dorylaimid nematodes having 1.5 to 5 mm body length in adults. Diagnostic features of ten important species are given in Table 2.

Biology

The duration of life cycle from egg to egg, is highly variable and under controlled conditions, it ranges from 3 weeks as in *X. index*[9], to 3 years as in *X. diversicaudatum*[10]. Individual females may live for more than 5 years. Life cycle of even the same species is also highly variable, depending on host, ecological factors and perhaps the biotype (Table 3). Reproduction is parthenogenetic in most of the species. Peak egg laying with 180-200 eggs per female, may occur in summer (April to May). Females lay eggs singly in soil[11]. Unlike tylenchid plant parasites, first moult takes place outside the egg and adults emerge after three more successive moults. In a recent study, the reproductive span of *X. index* has been recorded as 56 weeks, while the day degree requirement at a minimum threshold of 10°C for egg production varied from 24 to 26°C per egg on *Ficus carica* and 62 to 90° per egg on tomato[12].

Soil type is not a limiting factor in their distribution particularly for species like *X. americanum* and *X. index* which have been recorded from a number of geographical regions. However, some species like *X. brevicola*, *X. italiae*, *X. neovuittenezi*, etc., prefer sandy soils to heavy clay type[13]. The reverse is true for *X. diversicaudatum*[14]. Extremes of soil moisture (below 10 per cent and above 90 per cent) or wide fluctuations are not very conducive for nematode development. Temperature and pH requirements vary with the species. Generally, they prefer slightly acidic or almost neutral soil. The temperature range, however, show a wide divergence. *X. index* thrives well at 29.4°C while *X. americanum* multiplies well within the range of 20.0 to 24.0°C[15]. These nematodes have the capacity for overwintering, under European conditions, without any detriment to viability[16]. Some species are known to survive in soil for long periods without feeding. *X. diversicaudatum* can survive without host for 3 years[17] and *X. index* for 4-5 years[18].

The pathogenic level of the nematodes is highly variable and depends mainly on the species, host and the environment[19]. Twenty

TABLE 2

Diagnostic Characters of Some Important Species of Xiphinema

Diagnostic characters	Species name		
	X. americanum	*X. index*	*X. diversicaudatum*
Uterine differentiation	Uterus short (about one body width) & 'Z' differentiation absent	Uterus short (about one body width) & 'Z' differentiation absent	'Z' pseudo organ of 'B' type present
Vulval position	46-55%	37-43%	30-50%
Tail shape	Short & conoid with rounded terminus	Short bluntly rounded with prominent concavity on dorsal side and with a distinct digitate extension ventrally	Short bluntly rounded with prominent concavity on dorsal side and with a distinct digitate extension ventrally.
Coefficient 'C' value	1 to 2.5	1.5	1 to 1.5
Body length	1.4-2.0 mm	2.6-3.6 mm	3.5-4.5 mm

Spear length	101-123 um	183-213 um	200-250 um
Outline of anterior part of body	Weak to conspicuous constriction may be present	Without any constriction; may not be progressively attenuated	Without any constriction; may not be progressively attenuated
Habitus	Body spiral shaped	Open 'C' shaped	Open 'C' or 'J' shaped or hook shaped

Diagnostic characters	Species name			
	X. brevicolle	*X. coxi*	*X. vuittenezi*	*X. mediterraneum*
(1)	(2)	(3)	(4)	(5)
Uterine differentiation	Uterus short, about one body width & 'Z' differentiation absent	Z' pseudo organ or 'B' type present	Uterus short & 'Z' differentiation absent	Uterus short & 'Z' differentiation absent
Vulval position	50%	40-50%	40-50%	50%
Tail shape	Short and conoid with rounded terminus	Short conical digitate	Short bluntly rounded with concavity on dorsal side	Regularly short conical

(*Contd.*)

(Contd.)

(1)	(2)	(3)	(4)	(5)
Coefficient 'C' value	1.0	1.2.5	1.5	1-2.5
Body length	1.5-2.5 mm	2.5-4.5 mm	2.5-4.5 mm	1.5-2.5 mm
Spear length	150-200 μm	150-250 μm	200-250 μm	150-200 μm
Outline of anterior part of body	May be with weak constriction	May be with weak constriction	May be with weak constriction	With conspicuous constriction
Habitus	'C' or spiral shaped	Hook shaped, 'C' or 'J' shaped	Hook shaped, 'C' or 'J' shaped	Body spiral shaped

Diagnostic characters	Species name		
	X. insigne	*X. basiri*	*X. ifacolum*
Uterine	Uterus short and 'Z' differentiation absent	'Z' pseudo organ or 'B' type present	'Z' organ present
Vulval position	30-40%	40-50%	40-50%

Tail shape	Long conical	Short, conical, digitate in adults, while regularly short and unical in fourth stage juveniles	Regularly short conical
Coefficient 'C' value	2.5-5	1-2.5	1.5-2.5
Body length	1.5-2.5 mm	2.5-3.5 mm	2.5-4.5 mm
Spear length	200 μm	150-200 μm	150-200 μm
Outline of anterior part of body	Slight to conspicuous co-striction may be present	Slight constriction may be present	Slight constriction may be present
Habitus	Body slightly curved	Body from curved 'C' shape to spiral shape	Body curvature from 'C' shape to spiral shape

TABLE 3
Duration of life cycle of *Xiphinema* spp.

Nematode	Host	Duration
X. americanum	Cottonwood	1 year
X. brevicolle	Grapevine	4-7 months at 20-23°C
X. diversicaudatum	Orchards in South East England	3 years
X. index	Grapevine	21-27 days at 24°C
X. vuittenezi	Orchards in South East England	2 years

to eighty females of *X. ifacolum* per pot have been reported to reduce significantly height of black pepper seedlings[20]. With *X. vuittenezi*, 300-400 nematodes/1000 ml soil are required to cause 40 per cent reduction in root mass of several fruit trees[21], and 600-800 nematodes of this species as well as *X. mediterraneum* for causing severe symptoms on apple and pears[22].

Similarly, top and root weight, inflorescence number and fruit size of grapes are reported to be damaged extensively by 500 *X. index*/pot[23,24]. It is quite apparent that population levels are highly critical in determining the extent of damage. It has been clearly exemplified in the case of *X. diversicaudatum* which caused 650 per cent damage to rose at 500 nematodes and completely arrested growth at 10,000 nematode level[25].

Symptoms

These nematodes are ectoparasites, mostly feeding at or near the root tips. With the help of labelled P^{32}, it has been demonstrated that the nematodes (*X. vuittenezi*) suck the plant sap[26]. No specific above ground symptom is evident following feeding except stunting and growth decline symptoms that are common to many soil borne pathogen. Reduced root system, lesion formation, disintegration and decay of cortical tissue and discolouration are the general symptoms of infestation. Prominent terminal or subterminal swellings on roots, the symptom described as 'curly tip' or 'Fish hook galls' are often caused by *X. diversicauaatum* on rose, plum, celery, and strawberry ; by *X. bakeri* on forest tree saplings and by *X. index* on grapevine. The long stylet penetrates deep into the plant tissue, even upto vascular elements. In older and thicker roots, however, stylet penetration is limited to cortical parenchymal layers and the feeding duration is relatively longer compared to feeding at root tips[27]. Proteinaceous secretions by the nematodes[28] are suspected to be the cause of several histological changes in the host plants. A layer of phellogen tissue, 3-5 cells beneath the directly affected cells, may be formed[19]. Roots may be swollen, 7 to 9 hrs after odontostyle insertion in transition zone between root apex and zone of cell elongation. Production of dark suberized cells, reduction of apical meristems, both in size and activity, hypertrophy and hyperplasia of cortical cells occur following nematode feeding. Mitosis without cytokinesis may lead to multinucleate condition (2-8 nuclei) within 48 hrs. Cell organelles are rich in cytoplasm and nuclear envelope is invaginated[29–33].

These nematodes have shot into prominence due to their ability to transmit virus diseases. At least 9 species namely, *X. index* (grapevine fan leaf, grapevine banding, grapevine chrome mosaic and arabis mosaic); *X. americanum* (peach rosette mosaic, cherry raspleaf, tobacco ring spot and tomato ring spot) ; *X. diversicaudatum* (arabis mosaic, brome mosaic, carnation ring spot, cherry leaf roll, strawberry latent ring spot) ; *X. coxi* (arabis mosaic, brome mosaic, cherry leaf roll, strawberry latent ring spot and tobacco ring spot) ; *X. bakeri* (arabis mosaic) ; *X. vuittenezi* (cherry leaf roll) ; *X. basiri* (cowpea mosaic) ; *X. italiae* (grapevine fan leaf) and *X. brevicolle* (tomato ring spot) are implicated in transmitting at least 15 virus diseases. Informations on nematode virus interactions involving nematode feeding and acquisition of virus, transmission efficiency, specifically of virus retention and mechanism of virus transmission have been recently reviewed comprehensively[8,34]. These nematodes transmit NEPO viruses only. Virus acquisition period may take less than 5 minutes. Transmission efficiency is very high and single nematode may acquire and successfully transmit virus disease. Virus particles are adsorbed on the cuticle lining of the lumen of odontophore and oesophagus. However, the period of nematode feeding required to transmit virus disease may vary from few days to few months. Information on the effects of virus on nematode and vice versa is very limited. Fan leaf virus adversely affects osmoregulatory mechanism of *X. index* and even the enlarged nuclei and higher RNA content are attributed to the virus presence in nematode[35]. Viruliferous nematodes may reproduce better and survive longer in fallow soil[36].

Bacterial symbionts[37], rickettsias (intra-cellular bacterium like micro organism)[38], pathogenic bacteria[39], and micro-sporidia[40] are reported to be associated with dagger nematodes, though their role is little understood.

Genera *Longidorus* and *Paralongidorus*, the Needle Nematodes

Members of this group have gone through several taxonomical changes since the genus *Longidorus* was first erected in 1934. The two main genera, which contain the plant parasitic nematodes, are *Longidorus* and *Paralongidorus*. These are also commonly referred to as 'Needle nematodes' because of their long and thin needle like stylet. The well known species in the genus *Longidorus* are : *L. attenuatus*, *L. goodeyi*, *L. caespiticola*, *L. macrosoma*, *L. profundorum*, *L. vineacola* and *L. laevicapitatus*. These are also fairly widespread[41]. There are about 35 more species described under this genus but there is not much information known about them. In the allied genus, *Paralongidorus*, *P. maximus* and *P. sali* are the most important ones out of about 24 species described so far[41].

The body size is highly variable and ranges between 2 and 12 mm. The longest (12.35 mm) plant parasitic nematode is *Paralongidorus maximus*. Lip shape is variable, either continuous or with slight pre-labial constriction or with subacute shape with abrupt tapering at the anterior end.

Tail shape may vary from bluntly rounded to conoid with rounded terminus or conoid with more or less pointed terminus. Amphid opening is a transverse slit in *Paralongidorus*, while in

Longidorus it is a minute pore and the pouch is very large with or without bilobed base.

Life History

Life cycle is generally shorter than the dagger nematodes. However, *L. macrosoma* has a life span of 3 to 5 years and may take 2 years from egg to adult stage only[10]. Reproduction is parthenogenetic (meiotic) or amphimictic. Temperature is a major factor influencing duration of life cycle. Nematode development is arrested below 10°C. The life cycle is extended below 16°C and shortened between 16° and 28°C. Generally, the optimum temperature for development is 25°C[42,43]. However, it may vary with the nematode species also. For instance, for *L. elongatus* reproduction optimum is 25°C while for *L. africanus* it is 30°C. Further, the latter does not reproduce below 24°C[44].

The soil depth at which these nematodes are generally encountered depends largely on host root distribution, soil type and nematode species. *L. leptocephalus*, has been reported to be uniformly distributed upto a depth of 60 cm, with barley as the host; while with potato as the host, it is not encountered upto 20 cm and is distributed only between 40-60 cm[45].

These nematodes can survive freezing temperatures, but high soil moisture lessens their tolerance. The third and fourth larval stages as well as the eggs have the capacity to over-winter[44,45]. Generally, the nematodes prefer light sandy soils and are scarcely recorded from altitudes exceeding 800 ft[45,46].

Host Range and Pathogenesis

Generally, these nematodes occur around herbaceous perennials or herbaceous annuals. Forest tree seedlings, coniferous trees, vegetables and cereals are frequently associated with needle nematodes. Informations on pathogenic levels and crop losses are scarce, though circumstantial evidence and field associations reveal their economic importance to several crops. Pathogenicity of *L. elongatus* has been demonstrated on buck wheat and strawberry[47–49]. *L. africanus* reportedly causes 18 per cent reduction in weight of grapevine within six months[19]. A disease of sugarbeet, known as Docking disease, is found associated with high populations of *L. elongatus*, *L. atteruatus*, *L. caespiticola* and *L. leptocephalus*[50]. In this case, the tip of growing sugarbeet tap root may be killed and lateral roots are shortened with swollen tips leading to severe stunting in plants. Nematode affected areas in field appear as patches with nitrogen or manganese deficiency. Docking disorder of sugarbeet is confined to alkaline sandy soils of low organic contents[51]. *L. breviannulatus* is commonly associated with maize in U.S.A., causing seedling mortality at a level of 100 nematodes/100 cc of soil and is known to cause yield losses upto 62 per cent[52]. *L. elongatus* has been reported to cause considerable damage to strawberry, raspberry and grapes, pear and peppermint[53–55]. Vegetable crops such as bean and cucumber are attacked by *Paralongidorus maximus*, potato by *L. leptocephalus*, onion and celery by *P. maximus* and *L. vineacola*, carrot by *P. maximus*, *L. elongatus* and *L. vineacola* and lettuce by *P maximus*, *L. africanus* and *L. macrosoma.* Cereal crops such as barley have been recorded to be affected by *L. caespiticola*, oats by *L. cohni*, and sorghum by *L. africanus*. Ornamental crops such as gladiolus have been found attacked by *P. maximus* and rose by *L. macrosoma.* Similarly, several pasture legumes and turf grasses including alfalfa as also tobacco, cotton and several forest tree seedlings have been recorded to be attacked by one or more species of these nematodes[56]. Some of these are of economic importance in restricted areas.

These nematodes are typical root tip feeders, causing galls on several host plants. Meristem is inactivated. Feeding period may extend from few minutes to 4 days[57]. Irreversible root growth, cessation and swelling behind elongation zone are evident within 24 hrs[41]. Gall formation may be due to cortical hyperplasia alone with dense cytoplasm and enlarged nucleus near necrotic cells, as in celery[58] or due to both hypertrophy and hyperplasia as in burmary gold[43]. It has been estimated that there is increased activity of amino acids, carbohydrates, nucleic acids, phenols and peroxidase in galls formed by needle nematodes[59–61]. Histopathological studies of galls produced on *Chenopodium quinoa* by *L. apulus* have revealed that the nematode feeding led to multinucleate (6-8 nuclei) condition, increased cytoplasmic density, mitochondria and protoplastids[62].

As in the case of dagger nematodes, members of both the

genera, *Longidorus* and *Paralongidorus*, have the ability to transmit plant viruses. *L. macrosoma* transmits brome mosaic, carnation ring spot, pear storey pit, prunus necrotic ring spot and raspberry ring spot viruses ; *L. caespiticola* transmits arabis mosaic and raspberry ring spot virus. The tomato black ring, carnation ring spot, raspberry ring spot and tomato black ring, viruses can be transmitted by *L. attenuatus*, *L. elongatus* and *L. leptocephalus*. *L. apulus* (artichoke italian latent) *L. euonymus* (Euonymus mosaic) and *L. martini* (mulberry ring spot) are known also to transmit virus diseases. *Paralongidorus maximus* is implicated in transmitting arabis mosaic virus.

Informations on involvement with other microorganisms are very limited. *L. elongatus* in combination with *Helminthosporium sativum* has been reported to cause synergistic damage to spring wheat[63].

Genera *Trichodorus* and *Paratrichodorus*, The Stubby Root Nematodes

Trichodorus primitivus is the first stubby root nematode described, as early as 1880[64] under the name *Dorylaimus primitivus*. *Trichodorus* attained its generic status in 1913[6]. Among the known trichodorid nematodes, at least six species of *Trichodorus* namely, *T. primitivus*, *T. viruliferus*, *T. similis*, *T. proximus*, *T. sparsus* and *T. cylindricus* and eight of *Paratrichodorus* namely, *P. pachydermus*, *P. porosus*, *P. teres*, *P. nanus*, *P. allius*, *P. minor* (syn. *T. chrtstie* and *P. christiei*), *P. lobatus* and *P. anemones* are involved in causing considerable damage to various crops of economic significance. These are relatively short, 0.5 to 1.5 mm long, compared to dagger and needle nematodes.

Life History

Informations on life cycles of stubby root nematodes are scanty and incomplete. Hatching period varies with species and temperature. *P. minor* (=*P. christiei*) takes about 100-120 hrs for hatching[65] while *T. similis* takes about 12 days at 25°C which is extended at low temperature (10°C) even up to 65 days[66]. The time required for development of second-stage larva (*P. allius*) to adult is 22 days at 20°C[67]. *T. viruliferus* completes its life cycle in 45 days under laboratory conditions, and in 7-8 weeks on apple. Other species, for example, *T. primitivus* and *T. similis*, might also take roughly about the same time to complete their life cycles[68]. *P. minor* takes about 16-22 days on tomato and maize depending upon temperature to complete their life cycle[69].

Stubby root nematodes are very sensitive to various ecological parameters and need to be handled very carefully while sampling and extraction. These nematodes prefer light soils having 80 per cent or more sand and low in organic matter content[70]. Lack of moisture or dessication adversely affects nematode population and their numbers are highly correlated with rainfall[71]. Nematode movement is best in 200 to 400 μm fraction with soil pore space half full of water. Recovery of nematodes is high with increased diameter (10 cm) of sampling cores and minimum disturbance of samples after collection[72]. The nematodes are highly susceptible to copper ions or dissolved copper salts[73],[74] and are known to occur in soils having pH ranging from 4.2 to 8.1[75]. Due to the nematode's susceptibility to dessication, they are usually scarce in upper 5 cm soil depth. Optimum temperature for growth, reproduction and survival of stubby root nematodes is about 24°C±2°C[76]. The development is inhibited at 35°C.

Host Range and Pathogenesis

They are polyphagous nematodes and attack mainly vegetable and fruit crops, but occasionally cereals also[77–80]. Sugarbeet and potato are frequently attacked by one or more species (*T. cylindricus*, *T. primitivus*, *T. proximus*, *T. sparsus*, *T. viruliferus*, *P. allius*, *P. anemones*, *P. minor*, *P. nanus*, *P. pachydermus* and *P. teres*). Fruit crops such as apple, peach, pear, plum, walnut, strawberry, blue berry, citrus, grapevine, etc., are also susceptible to stubby root nematodes (*T. viruliferus*, *T. similis*, *T. primitius*, *P. minor* and *P. porosus*). *T. viruliferous*, *T. primitivus*, *P. minor* and *P. anemones* attack some of the cereal crops (wheat, maize, barley rye, etc.). *P. minor* has a very wide host range which includes several vegetable crops, sugarcane and tobacco. The latter two crops are attacked by *P. lobatus* also. Physiological races having differences in host preference might exist in *P. minor*[67].

The nematodes aggregate around the growing root tips and feed in the zone of elongation. Feeding is normally restricted to epidermal cells and root hairs, while gregarious feeding may extend upto cortical cells. Feeding duration is usually confined to few minutes in individual cells, except *T. proximus* which may feed up to 170 minutes. Decline in root growth, discolouration, superficial cracks on roots, root growth cessation and stubby root formation may occur due to nematode feeding, leading to stunted plant growth.

Cytoplasm and some times contents of nucleus are ingested by the nematodes. Nucleus gets disorganised and the cytoplasm appears coagulated. Epidermal and outer cortical cells may collapse with extensive feeding[58]. *P. minor* may induce cell division in pericycle resulting in flattened and abnormally enlarged rootlets[77,81].

Informations on pathogenic levels of the nematode affecting plant growth are limited. *P. minor*, at high levels (2000 nematodes/litre of soil), though did not cause any quantitative damage to cotton, it is reported to cause significant reduction in seedling weight at low level (160-533 *P. minor*/litre). Pine seedling weight and height are reported to be significantly reduced by *P. minor* at levels of 500 or 5000 nematodes, six months after inoculation. Stubby root nematodes are involved in docking disorder of sugarbeet wherein it has been estimated that for every 500 nematodes, there is a loss of 12 per cent in roots. Further, there existed a significant negative regression between numbers of stubby root nematodes and sugarbeet roots.[82]

At least twelve species are implicated in transmission of virus diseases. The species involved and the virus diseases alongwith the strains transmitted are : *P. minor* (tobacco rattle ; California and Wisconsin), *P. pachydermus* (tobacco rattle ; Dutch and English ; pea early browning Dutch and English). *P. teres* (tobacco rattle English and Oregon ; pea early browning : Dutch) *P. anemones* (Pea early browning, English). *P. porosus* (tobacco rattle, California), *P. allius* (tobacco rattle, California and Oregon), *P. nanus* (tobacco rattle, Dutch and English), *T. viruliferus* (Pea early browning, English and

Italian No. 6 ; tobacco rattle, Dutch). *T. primitivus* (tobacco rattle, English, German and Scottish), *T. cylindricus* (tobacco rattle, Dutch), *T. hooperi* (tobacco rattle) and *T. similis* (tobacco rattle, Dutch ; gladiolus notch leaf). *P. minor* is reported to interact positively with *Fusarium moniliforme* on sugarcane root growth[83,84].

Management

Though several fumigant and non-fumigant nematicides are effective in managing nematode populations as such, none of them suits to the standards of achieving 100 per cent control of nematode vectors of plant viruses. The reasons may be due to non-availability of effective ovicidal nematicides and also because they are well protected in soil medium, at times several meters deep alongwith the perennial tree roots, where nematicidal application is cumbersome and costly. However, dagger, needle and stubby root nematodes can be effectively managed better as pathogens than as virus vectors.

Fumigants like DD and EDB have been demonstrated to be quite effective in reducing nematode populations[85–87]. Granular nematicides like phenamiphos, ethoprop, aldicarb, etc., have now replaced the use of fumigants. Phenamiphos has proved effective in apples and grapes against dagger nematodes[88,89]. At 30 kg a.i./ha, it was also effective against stubby root nematode in potato. Ethoprop 10 G at 3.4 kg a.i./ha has been recorded to be as effective as phenamiphos in reducing populations of *Paratrichodorus minor* and alleviating the corky ring spot disease of potato. The carbamate group of chemicals have been successfully employed in management of these nematodes[90–94].

Among the non-chemical methods, growing non-host/enemy/trap crops, in rotations or use of resistant/tolerant varieties have been attempted. *Crotolaria spectabilis*, as a green manure crop not only improves nitrogen status of the soil but also reduces the stubby root nematode population[93]. Antagonistic crops such as *Tagates erecta* and *Zinnia elegans* have been reported to suppress the trichodorid nematode populations. Aqueous leaf extract of *T. patula* has been observed also to have nematostatic properties against *X. basiri*[94]. Efforts to locate resistant and tolerant varieties against dagger nematodes and grape fan leaf virus in grapevine are successful and can be effectively included in nematode management schedules. Muscadine grapes (*Vitis rotundifolia*) are resistant to grape fan leaf virus transmitted by *X. index*[95], while sever l Californian hybrid root stocks are resistant and some of them are immune to *X. index*[96]. Nematode management, particularly with reference to dagger, needle and stubby root nematodes, by altering host nutrition, has received little attention. However, it has been observed that lime deficiency increases *P. minor* population and this indicates possibility of reducing trichodorid nematodes by addition of lime.

Excessive or continuous use of either nematicides or resistant varieties lead to ecological imbalances either by killing non-target organisms or by promoting development of new races of pathogens. Hence, location specific management schedules need to be developed by combining judiciously chemical and non-chemical methods for effective pest management.

REFERENCES

1. Siddiqi, M.R., Hooper, D.J. and Khan, E., A new nematode genus *Paralongidorus* (Nematoda : Dorylaimoidea) with description of two new species and observations on *Paralongidorus citri* (Siddiqi, 1959) n. comb. *Nematologica*, 9, 7, 1963.

2. Siddiqi, M.R., Systematics of the genus *Trichodorus* Cobb, 1913 (Nematoda : Dorylaimida), with description of three new species. *Nematologica*, 19, 259, 1974.

3. Christie, J.R. and Perry, V.G., A root disease of plants caused by a nematode of the genus *Trichodorus*. *Science*, 113, 491, 1951.

4. Hewitt, W.B., Raski, D.J. and Goheen, A.C., Nematode vector of soil borne fan leaf virus of grapevines. *Phytopathology*, 48, 586, 1952.

5. Cobb, N.A., Nematodes, mostly Australian and Fijian. *Mac Leay Mem. Vol., Linn. Socs. N.S.W.*, 252, 1893.

6. Cobb, N.A., New nematode genera found inhabiting fresh

water and non-brackish soils. *J. Wash' Acad Sci.*, 3, 432, 1913.

7. Luc, M., Taxonomy of *Xiphinema* Cobb, 1913. In : Nematode Vectors of Plant Viruses (Eds. Lamberti, F., Taylor, C.E. and Seinhorst, J.W.). *Plenum Press*, London, 39, 1975.

8. Taylor, C.E. and Brown, D.J.F., Nematode virus interactions. In: Plant Parasitic Nematodes, Vol. III (Eds. Zuckerman, B.M. Mai, W.F. and Rohde, R.A.). *Academic Press* Inc. (London) Ltd., London, Chap. 11, 1981.

9. Radewald, J.D. and Raski, D.J., A study of life cycle of *Xiphinema index*. *Phytopathology*, 52, 748, 1962.

10. Flegg, J.J.M., Life cycle studies of some *Xiphinema* and *Longidorus* species in South-Eastern England. *Nematologica*, 14, 197, 1968.

11. Brown, D.J.F. and Coire, M.I., The total reproductive capacity and longevity of individual female, *Xiphinema diversicaudatum* (Nematoda : Dorylaimida). *Nema tologia Mediterr.*, 11, 87, 1983.

12. Brown, D.J.F. and Coire, M.I., The reproductive capacity and longevity of *Xiphinema index* (Nematoda : Dorylaimida) from three populations on selected host plants. *Revue Nematol.*. 8, 171, 1985.

13. Cohn, E., The occurence and distribution of species of *Xiphinema* and *Longidorus* in Israel *Nematologica*, 15, 179, 1969.

14. Harrison, B.D. and Winslow, R.D., Laboratory and field studies on the relation of arabis mosaic virus to its nematode vector *Xiphinema diversicaudatum* (Micoletzky), *Ann. Appl. Biol.*, 49, 521, 1961.

15. Lownsbery, B.F. and Maggenti, A.R., Some effects of soil temperature and soil moisture on population levels of *Xiphinema americanum*. *Phytopathology*, 53, 667, 1963.

16. Flegg, J.J.M., Once yearly reproduction in *Xiphinema vuittenezi. Nature*, London, 212, 741, 1966.

17. Pitcher, R.S., Siddiqi, M.R. and Brown, D.J.F., C.I.H. Description of Plant Parasitic Nematodes, *Xiphinema diversicaudatum*. Commonwealth Agricultural Bureaux, London, Set 4, No. 60, 1974.

18. Raski, D.J., Hewitt, W.B., Goheen, A.C., Taylor, C.E. and Taylor, R.H., Survival of *Xiphinema index* and reservoirs of fan leaf virus in fallowed vineyeard soil. *Nematologica*, 11, 349, 1965.

19. Cohn, E. and Orion, D., The pathological effect of representative *Xiphinema* and *Longidorus* species on selected host plants. *Nematologica*, 16, 423, 1970.

20. Lamberti, F., Rohini, H.M. and Ekanayake, K., Effect of some plant parasitic nematodes on the growth of black pepper in Sri Lanka. *Plant Protection Bulletin*, 31, 163, 1983.

21. Jenser, G. add Simon, I., A *Xiphinema vuittenezi* fitopato genitasanak, vizsgalata gyumoicsfock gyokeren. *Agrartudomanyi Kozylemenyek*, 39, 102, 1980.

22. Tiev, R.A., Longidorids of Fruit Trees in the Kabardino Balkar ASSR (ED. K.I. Skryabin). *Byulletin vesesoyuznogo Institute gel mintologli*, No. 31, 49, 1981.

23. Krikpatrick, J.D., Van Gundy, S.D. and Martin, J.P., Effects of *Xiphinema index* on growth and abscission in carignane grape, *Vitis vinifera*. *Nematologica*, 11, 41, 1965.

24. Boublas, D., Pistre, R., Dalmasso, A. and Bongiovani, M., Aspect des atteintes sur racines de vigne du nematode *Xiphinema index* vecteur du courtnoue de la *vigne*, *Progress agric. Vitic.*, 171, 1, 1971.

25. Schindler, A.F., Parasitism and pathogenecity of *Xiphinema diversicaudatum*, an ectoparasitic nematode. *Nematologica*, 2, 25, 1957.

26. Jenser, G. and Erdelyi, G., A *Xiphinema vuittenezi* mint feltetelezett virus Vektor Taplallk felvetelenek Vizsgalata forz for isotappal jelzett novenyeken. *Agrartudomanyi Kozlemenyek*, 39, 101, 1980.

27. Cohn, E., Relations between *Xiphinema* and *Longidorus* and their host plants. In : Nematode Vectors of Plant Viruses (Eds. Lamberti, F., Taylor, C.E. and Seinhorst, J.W.). *Plenum Press*, London, 365 pp., 1975.

28. Poehling, H.M. and Wyss, U., Microanalysis of proteins in the aseptic host parasite system *Ficus carica-Xiphinena index. Nematologica*, 26, 230, 1980.

29. Davis, R.A. and Jenkins, W.R., Nematodes associated with roses and the root injury caused by *Meloidogyne hapla* Chitwood, 1949, *Xiphinema diversicaudatum* (Micoletzky, 1927) Thorne, 1939 and *Helicotylenchus nanus* Steiner, 1945. *Bull. Md. agric. Exp. Stn.* A, 106, 1, 1960.

30. Radewald, J.D. and Raski, D.J., Studies on the host range and pathogenicity of *Xiphinema index* (Abstr.). *Phytopathology*, 52, 748, 1962.

31. Weischer, B. and Wyss, U., Development, histology and ultrastructure of root tip galls induced by the ectoparasitic nematode, *Xiphinema index*. Bulletin de la societe Botanique de France, *Actualites Botaneque*, 127, 67, 1980.

32. Bleve-Zacheo, T. and Zacheo, G., Early stage of disease in fig roots induced by *Xiphinema index. Nematologia Mediterr*, 11, 175, 1983.

33. Johnsen, H.J., Nematode pests of vegetable and related crops. In : Economic Nematology (Ed. Webster, J.M.). Academic Press Inc. (London) Ltd., London, 377, 1972.

34. Das, S.N., Transmission of plant viruses by nematode vectors. In : Plant Parasitic Nematodes of India (Eds. Swarup, G. and Dasgupta, D.R.). Indian Agricultural Research Institute, New Delhi, 1980, 412.

35. Roggen, D.R., On the morphology of *Xiphinema index* reared on grape fan leaf virus infected grapes. *Nematologica*, 12, 287, 1966.

36. Das, S.N. and Raski, D.J., Effect of grapevine fan leaf virus on the reproduction and survival of its nematode vector, *Xiphinema index* Thorne and Allen. *J. Nematol.*, 1, 107, 1969.

37. Mankau, R., Microbial Control of Nematodes. In Plant Parasitic Nematodes, Vol. III (Eds. Zuckerman, B.M. Mai, W.F. and Rohde, R.A.). Academic Press Inc., New York, Chap. 19, 1981.

38. Rumbos, I., Sikora, R.A. and Neinhuas, F., Rickettsia like organisms in *Xiphinema index* (Thorne) Allen found associated with yellow disease of grapevines. *Z. Pflanzenker Pflanzenschutz.*, 84, 240, 1977.

39. Adams, R.E. and Fichenmuller, J.J., A bacterial infection of *Xiphinema americanum*. *Phytopathology*, 53, 745, 1963.

40. Morone di Lucia, M.R. and Grimaldi, de Zeo S., Presenza di microsporidi in gonadi di *Xiphinema mediterraneum*. *Nematologia Mediterr.*, 1, 66, 1973.

41. Lamberti, F., Taxonomy of *Longidorus* (Micoletzky) Filipjev and *Paralongidorus* Siddiqi, Hooper and Khan. In : Nematode Vectors of Plant Viruses (Eds. Lamberti, F., Taylor, C.E. and Seinhorst, J.W.). *Plenum Press*, London, 71, 1975.

42. Cohn, E. and Mordechai, M., The influence of some environmental and cultural conditions on rearing populations of *Xiphinema* and *Longidorus*. *Nematologica*, 16, 85, 1970.

43. Wyss, U., Untersuchungen Zur populations dynamik von *Longidorus elongatus*. *Nematologica*, 16, 74, 1970.

44. Lamberti, F., Effect of temperature on the reproduction rate of *Longidorus africanus*. *Plant Dis. Reptr.*, 53, 559, 1969.

45. Boag, B. and Brown, D.J.F., C.I.H. descriptions of Plant Parasitic Nematodes, *Longidorus leptocephalus. Commonwealth Agricultural Bureaux*, London, Set. 6, No. 88, 1976.

46. Lamberti, F., The effect of cropping on the population levels of *Longidorus africanus. Plant Dis. Reptr.*, 52, 748, 1968.

47. Semkina, A.I., Effect of the nematode, *Longidorus elongatus* (De Man, 1876) on the yield of buck wheat. In : *Helminthological Abstracts*, Series B, 1981.

48. Seinhorst, J.W., *Longidorus elongatus* on *Fragaria vesca. Nematologica*, 12, 275, 1966.

39. Sharma, R.D., Direct damage to strawberry by *Longidorus elongatus. Meded Rijkslandb. Hogesch. Gent.*, 30, 1437, 1965.

50. Dunning, R.A. and Cooke, D.A., Docking Disorder of Sugarbeet. Min. Agric. Fish and Foods, Middle sex, England, 1972.

51. Steele, A.E., Nematode parasites of sugarbeet. In : Plant and Insect Nematodes (Ed. Nickle, W.R.). Marcel Dekker Inc., New York, Chap, 14, 1984.

52. Malek, R.B., Norton, D.C., Jacob Sen, B.J. and Acosta, N., A new corn disease caused by *Longidorus breviannulatus* in the mid west. *Plant Dis.*, 64, 1110, 1980.

53. Raski, D.J. and Krusberg, L.R., Nematode Parasites of Grapes and Other Small Fruits. In : Plant and Insect Nematodes (Ed. Nickle, W.R.). *Marcel Dekker, Inc.*, New York, Chap. 83, 1984.

54. Wehunt, E.J., Nematode Parasites of Peach and Other Tree Crops. In Plant and Insect Nematodes (Ed. Nickle, W.R.). *Marcel Dekker, Inc.*, New York, Chap. 12, 1984.

55. Jensen, H.J. and Horner, C.E., A decline of peppermint caused by an ectoparasitic nematode, *Longidorus sylphus. Phytopathology*, 46, 637, 1956.

56. Eriksson, K.B., Nematode diseases of pasture legumes and turf grasses. In : Economic Nematology (Ed. Webster, J.M.). *Academic Press Inc.* (London) Ltd., London, Chapt. 4, 1972.

57. Fritzsche, R. and Hofferek, H., Nahrung Saufnahme und Nahrpflanzenkereis von *Longidorus macrosoma*, Hooper, *Arch. Pflansenschutz.* 5, 423, 1969.

58. Wyss, U., Feeding of *Trichodorus*, *Longidorus* and *Xiphinema.* In : Nematode Vectors of Plant Viruses (Eds. Lamberti, F., Taylor, C.E. and Seinhorst, J.W,). *Plenum Press*, London, 203, 1975.

59. Epstein, E., Biochemical changes in terminal root galls caused by an ectoparasitic nematode, *Longidorus africanus* ; phenols, carbohydrates and cytokinins. *J. Nematol.*, 4, 246, 1972.

60. Epstein, E., Biochemical changes in terminal root galls caused by an ectoparasitic nematode, *Longidorus africanus* : Nucleic acids. *J. Nematol.*, 6, 48, 1974.

61. Epstein, E. and Cohn, E., Biochemical changes in terminal root galls caused by an ectoparasitic nemotode, *Longidorus africanus* : amino acids. *J. Nematol.*, 3, 334, 1971.

62. Blevezacheo, T., Yeves, M.F.A. and Zacheo, G., Development of galls induced in *Chenopodium guinoa* by *Longidorus apulus. Nematologia Mediterr.*, 12, 129, 1984.

63. Semkida, A.I., The nematode, *Longidorus elongatus* (de Man, 1876) and its effect on the harvest of spring wheat from *Helminthological Abstracts*, 52, Series B, Commonwealth Agril. Bureaux, London, No. 79, 1982.

64. De Man, J.G., Die einheimischen, Frei in der neinen Erde und im sussen wasser lebenden, Nematoden. *Tijdschrned dierk*, ver., 5, 1, 1880.

65. Bird, G.W., Goodman, R.M. and Mai, W.F., Observation on the embryogenesis of *Trichodorus christiei* (Nematoda : Diphtherophoroidea). *Can. J. Zool.*, 46, 292, 1968.

66. Wyss, U., C.I.H. description of Plant Parasitic Nematodes, *Trichodorus similis*. Commonwealth Agril. Bureaux, Set. 4, No. 59, 1974.

67. Ayala, A., Allen, M.W. and Noffsinger, E.M., Host range, biology and factors affecting survival and reproduction of the stubby root nematode. *J. Agr. Univ. P.R.*, 54, 341, 1970.

68. Pitcher, R.S. and McNamara, D.G., The affect of nutrition and season of year on the reproduction of *Trichodorus viruliferous*. *Nematologica*, 16, 99, 1770.

69. Heyns, J., C.I.H. descriptions of Plant Parasitic Nematodes, *Paratrichodorus christiei : Commonwealth Agril. Bureaux*, London, Set. No. 5, No. 69, 1975.

70. Cooke, D.A., The effect of plant parasitic nematodes, rainfall and other factors on Docking disorder of sugarbeet. *Plant Path.*, 22, 161, 1973.

71. Cooke, D.A. and Dryacott, A.P., The effects of soil fumigation and nitrogen fertilisers on nematodes and sugarbeet in sandy soils. *Ann. Appl. Biol.*, 69, 253, 1971.

72. Bor, N.A. and Kuper, K., Gevoeligheid van *Trichodorus teres* en *T. pachydermus* voor uitwendige involoeden ? *Meded. Rijks, Land Hogesch. Gent.*, 31, 609, 1966.

73. Van Gundy, S.D. and Thomason, I.J., Factors influencing storage life of *Hemicycliophora arenaria*, *Pratylenchus scribneri* and *Trichodorus christiei*. *Phytopathology*, 52, 366, 1962.

74. Hafkenscheid, H.H.M., Influence of Cu^{++} ions on *Trichodorus pachydermus* and an extraction method to obtain active specimens. *Nematologica*, 17, 535, 1971.

75. Van Hoof, H.A., Maal, D.Z. and Seinhorst, J.W., Viruses of the tobacco rattle virus group in Northern Italy : their vectors and serological relationships. *Neth. J. Pl. Path.*, 72, 253, 72, 253, 1966.

76. Malek, W.R., Jenkins. W.R. and Powers, E.M., Effect of temperature on growth and reproduction of *Criconemoides*

curvatum and *Trichodorus christiei. Nematologica*, 11, 41, 1965.

77. Standifer, M.S. and Perry, V.G., Some effects of sting and stubby root nematodes on grapefruit roots. *Phytopathology*, 50, 152, 1960.

78. Hooper, D.J. and Siddiqi, M.R., C.I.H. descriptions of Plant Parasitic Nematodes, *Trichodorus primitivus. Commonwealth Agril. Bureaux*, London, Set. 1, No. 15, 1972.

79. Hooper, D.J., C.I.H. description of Plant Parasitic Nematodes, *Trichodorus, viruliferous*, Commonwealth Agricultural Bureaux, London, 1976, Set No. 6, No. 86.

80. Allen, M.W., A review of the nematode genus *Trichodorus* with descriptions of ten new species. *Nematologica*, 2, 32, 1957.

81. Zuckermann, B.M., Parasitism and pathogenesis of the cultivated high bush blueberry by the stubby root nematode. *Phytopathology*, 52, 1017, 1962.

82. Brown. E.B. and Sykes, G.B., Studies on the relation between density of *Longidorus elongatus* and growth of sugarbeet with supplementary observations on *Trichodorus* spp, *Ann. of App. Biol.*, 68, 291, 1971.

83. Liu, L.J. ond Ayala, A., Pathogenicity of *Fusarium moniliforme* and *F. roseum* and their interaction with *Trichodorus christiei* on sugarcane in Puerto Rico. *Phytopathology*, 60, 1540, 1970.

84. Liu, L.J. and Ayala, A., Pathogenicity of *Fusarium* spp. and relation with *Trichodorus christiei* on sugarcane in Puerto Rico. *Nematropica*, 1, 7, 1971.

85. Alphey, I.J.W., Copper, J.I. and Harrison, B.D., Chemical control of trichodorids and of transmission of tobocco rattle virus. In : Nematode Vectors of Plant Viruses (Eds. Lamberti. F., Taylor, C.E. and Seinhorst, J.W.). Plenum Press, London, 441. 1971

86. Minton, N.A. and Parker, M.B., Response of soybeans and nematodes to all planting applications of fumigant and non-fumigant nematicides. *Nematropica*, 10, 31, 1980.

87. Smith, P.C., Control of *Xiphinema* in South Africa. In : Nematode Vectors of Plan Viruses (Eds. Lamberti, F., Taylor, C.E. and Seinhorst, J.W.). Plenum Press, London, 449, 1975.

88. Niles, R.K., Yoder, K.S. and Elliott, A.P., The effect of nema cur 35 on plant parasitic nematodes and apple growth and yield in Virginia, 1982. *Fuegicide and Nematicide Tests*, 38, 1, 1983.

89. Hafez, S.L., Raski, D.J. and Lear, B., Action of systemic nematicides in control of *Xiphinema index* on grape. *J. Nematol.*, 13, 24, 1981.

90. Weingartner, D.P., Shumaker, H.R. and Smart, G.C., Chemical control of croky ring spot disease on potatoes in Florida, U.S.A. In : Nematode Vectors of Plant Viruses (Eds. Lamberti, F., Taylor, C.E. and Seinhorst, J.W.). Plenum Press, London, 443, 1975.

91. Pinkerton, J.N. and Jensen, H.J., Chemical control of *Longidorus elongatus* on peppermint with nonvolatile nematicides. *Plant Disease*, 67, 201, 1983.

92. Spaull, A.M. and Newton, P.G., Poor establishment and yield losses of cereals and grasses caused by trichodorid nematodes. In Proc. of Crop Protection in Northern Britain, Rothamsted Experimental Station, Harpenden, Herts, 49, 1984.

93. Thompson, J.A., Miller, R.W., Conrad, N. and Carpenter, A.S., Evaluation of post plant nematicides for control of the dagger nematode on apples. *Fungicide Nematicide Tests*, 39, 87, 1984.

94. Indra, R., Varma, M.K. and Yadav, B.S., Nematostatic properties of *Tagetes patula* L. aqueous leaf extract on

Xiphenama basiri Siddiqi, 1959. *Indian J. Nematol.*, 15, 195, 1985.

95. Bouquet, A. and Danglot, Y , Recherche de porte greffes de vigne registant a la transmission du virus du court none (GFV) par le nematode *Xiphinema index* Thorne and Allen. I. Application de la methode ELISA ala realization d'um test rapid de selection. *Agronomie*, 3, 957, 1983.

96. Harris, A.R., Resistance of some *Vitis* root stocks to *Xiphinema index. J. Nematol.*, 15, 405, 1983.

Index

M

N

O

P